Quantum Computing

by whurley and Floyd Smith

Quantum Computing For Dummies®

Published by: **John Wiley & Sons, Inc.,** 111 River Street, Hoboken, NJ 07030-5774, www.wiley.com

For general information on our other products and services, please contact our Customer Care Department within the U.S. at 877-762-2974, outside the U.S. at 317-572-3993, or fax 317-572-4002. For technical support, please visit https://hub.wiley.com/community/support/dummies.

Wiley publishes in a variety of print and electronic formats and by print-on-demand. Some material included with standard print versions of this book may not be included in e-books or in print-on-demand. If this book refers to media such as a CD or DVD that is not included in the version you purchased, you may download this material at http://booksupport.wiley.com. For more information about Wiley products, visit www.wiley.com.

Library of Congress Control Number: 2023942826

ISBN 978-1-119-93390-8 (pbk); ISBN 978-1-119-93391-5 (ebk); ISBN 978-1-119-93392-2 (ebk)

SKY10059609_110823

Contents at a Glance

Table of Contents

Introduction

Quantum computing is a lot like the sulfur-spewing hydrothermal vents at the bottom of the Mariana Trench: It's both hot and deep. (The vents in the Mariana Trench are also surrounded by strange life forms, which implies something not very nice about us, your authors.)

Quantum computing is hot because progress, activity, interest, and investment are all continuing to grow to unprecedented levels. And it's hot also because popular culture says so. Movies like the 2023 Oscar winner, *Everything Everywhere All at Once*, and the Netflix series *3 Body Problem* make quantum mechanics — the technology that underlies quantum computing — lively, interesting, and even funny.

And why is quantum computing deep? Because it directly applies quantum mechanics to some of the biggest problems that humanity faces today, such as drug discovery, lifespan enhancement, and climate change. And the workings of the quantum world challenge our understanding, sometimes even our grasp on, reality.

Entanglement is a core principle of quantum computing, and the 2022 Nobel Prize in Physics was awarded for work on entanglement. Yet no less a figure than Albert Einstein called entanglement "spooky action at a distance." Quantum computing depends on this "spooky action" and other quantum mechanical principles to deliver incredible results.

All this makes the two of us very happy to bring you *Quantum Computing For Dummies*. We hope this book will direct your attention toward the big picture where needed — and also show you when it's time to, as the expression says, "shut up and calculate." Or program. Or make hard-nosed business decisions about where to put your time, energy, and money.

About This Book

Quantum Computing For Dummies provides a clear and concise introduction to the terminology, technology, and techniques you need to use quantum computing to get a job, make business decisions, invest, or just introduce yourself to this fascinating new technology.

With this book as your guide, you'll learn about

- The core principles of quantum mechanics
- How quantum mechanics relates to quantum computing
- Where the technology is today and how it relates to your business and career
- How quantum-inspired computing, a transitional technology, is being used to solve advanced problems on today's computers
- The difference between quantum annealing, another transitional technology, and gate-based quantum computing — the most complete expression of quantum computing technology
- Which gate-based quantum computing modality to use, based on nearly a dozen types of qubits
- More than a dozen quantum algorithms that might help to solve problems that your business faces today
- Identifying which kind of speedup quantum computing might offer for the problems that affect your business today
- The steps needed to make your business ready for — or a leader in — quantum computing
- Choosing a cloud provider to help you get started with the technology, often for free
- Online courses on how to program quantum computers
- Signing onto a live portal and starting your own quantum computer programming journey
- Resources, online and offline, to further your knowledge or quantum computing
- University programs to help you take the next step toward a quantum computing career

You're going to be seeing news stories about progress in quantum computing, and you may find yourself discussing the technology with friends or at work. You may even consider taking your career in this new direction. All of which leads to a simple question: What can you actually do about quantum computing?

This book answers that question. It enables you to relate the computers we all use today — called classical computers — to this strange new beast, quantum computing. It shows you what you can do with quantum computers today, and what you will likely be able to do in the future. And it punctures many myths about quantum computing.

The book then goes on to show you how to get started in quantum computing at work, whether that's programming a quantum computing or using quantum computing in your business. And it helps you understand the technology trends so you can keep an eye on new opportunities as quantum computing moves forward.

Quantum computing is quite different than the classical computers you're accustomed to. This book can help you decide when to use each type of computing, now and in the future.

You can learn bits and pieces about quantum computing by reading news stories and poking around online. But with this book, you get a complete picture. You can decide whether you and your business need to get started with quantum computing today — and if so, you can roll up your sleeves and get to work.

Foolish Assumptions

Quantum Computing For Dummies is written for beginners, which means you might not even know for sure what quantum computing is. We don't assume that you know physics or mathematics beyond what common sense tells you. And we don't assume that you've programmed a computer (of any kind) before.

However, we do assume that you

- Have used a computer or a smartphone or both
- Have access to the internet so you can do further research on new topics
- Are interested in what quantum computing can do for your career, your business, or the world

Icons Used in This Book

If you've read other *Dummies* books, you know that they use icons in the margin to call attention to particularly important or useful ideas in the text. In this book, we use four of these icons.

TIP

The tip icon highlights expert shortcuts or simple ideas that can make life in the quantum computing field easier for you.

TECHNICAL STUFF

You could say that this whole book is technical stuff, but this icon highlights information that's particularly technical. You can re-read information highlighted with this information if you want to go deep, or skip it if you don't want to know all the details. It's entirely up to you!

REMEMBER

While we would like to think that every word in this book is unforgettable, we'd like you to carry some information to other areas of the book. We highlight these points so you can quickly find important takeaways that matter quite a bit in helping you get up to speed with quantum computing.

WARNING

Just as you would do if you were driving a car, slow down when you see a warning sign. It highlights an area where it's easy to make a mistake or develop a misunderstanding.

Beyond the Book

To get to the web page for this book, go to `www.dummies.com` and type *Quantum Computing For Dummies* in the Search box. The web page includes the book's cheat sheet, which defines basic terms, describes the different types of quantum computers, lists the most-used types of qubits, and links to some of the online classes about quantum computing. You'll also find bonus chapters with tech and business information.

Occasionally, the publishers provide updates to *Dummies* books. If this book has technical updates, you can find them here as well.

To see the free programming introduction mentioned in Chapter 12, go to `www.dummies.com/go/quantumcomputingfd`.

One of the book's authors, William Hurley (known as whurley), maintains a personal website with information about quantum computing and other topics of interest. You can visit it at `www.whurley.com`. whurley is also CEO of Strangeworks, a quantum computing company. You can visit Strangeworks at `www.strangeworks.com`.

Where to Go from Here

Part 1 introduces you to quantum computing and the most closely related technologies, quantum mechanics and classical computing (the computers you use today). Part 2 is a deep dive into the different ways you can do quantum computing today and the different qubit types used in gate-based quantum computing, the most advanced (and, so far, least developed) form of the technology. Part 3 is quite practical, if we can use that word to describe an almost mystical technology; it introduces you to quantum computing algorithms, quantum computer programming, and educational resources for learning more. And Part 4 is the famous *Dummies* "Part of Tens"; it gives you lists of key points mentioned in the book, covering both technical and business information.

You can read this book in any way you'd like. If you go straight from beginning to end, the book will take you on a journey from the basics of quantum computing and classical computing to in-depth knowledge of technical topics such as qubits and quantum computing algorithms. It will then show you how to access and program quantum computers and give you a variety of ways to move yourself, your career, and your organization forward with the technology.

At the same time, we understand that you might want to pop in and out of the book. You may want to go deep on topics that interest you and skim other sections. You can look at the topics that interest you in the order that we have them arranged in the book; in your own order, based on your own interests; or in no specific order at all. The book is full of cross-references and signposts to help you find all the information you need.

1 The Power of Quantum Computing

IN THIS PART . . .

Get a big-picture overview of quantum computing hardware and software and the current state of progress in the field.

Learn about early computing and classical computing, the approach that describes the many types of computers, from PCs to smartphones to Roombas, that people use in daily life today.

Investigate the roots of quantum computing, made up of discoveries in quantum mechanics which burst onto the scene in the early 1900s, led by figures such as Albert Einstein and Neils Bohr.

Find out how quantum mechanics was first used to develop early quantum computing technologies such as X-ray machines, television, and laser beams.

Come along as we follow quantum mechanics technology into the early days of quantum computing, including the first quantum computing algorithms and the first qubits.

Observe the emergence of quantum computing in the last few years as research and development, interest, and venture capital investment all rise quickly.

» Calculating the power of quantum computing

» Identifying quantum computing's advantages

» Taking stock of quantum computer types

» Overcoming barriers to new capabilities

Chapter 1
Quantum Computing Boot Camp

Picture flipping a coin in the air. As it's spinning, is it showing heads or tails? Well, you can't know the answer while the coin is spinning. Only when the coin lands and settles down does it display a definite result.

The uncertainty you see while the coin is spinning is like the uncertainty we capture and use in quantum computing. We put many processing elements — *qubits* — into a state of uncertainty. Then we program the qubits, run the program, and capture the results — just like when the coin lands.

Quantum computing is different from the fixed 0s and 1s, bits and bytes, used in today's devices. Quantum computing is based on quantum mechanics, a branch of physics that can be hard to comprehend. But the way in which quantum computing deals effectively with large degrees of uncertainty feels like the way we make many of the decisions we encounter in daily life.

Quantum computing is complementary to classical computing, the kind of computing we use today, not a replacement for it. By working with uncertainty, we can take on some of the biggest, most complex problems that humanity faces, in a new and powerful way. Quantum computing will solve problems for which today's computing falls short — problems in areas such as modeling the climate, drug discovery, financial optimization, and whether or not it's a good morning to launch a rocket.

Quantum computing is just getting started; many advanced quantum computers run only for a fraction of a second at a time. However, steady progress is being made. Even now, at this early stage, quantum computing is inspiring us to, as a sage once said, "think different" about the way we use existing computing capabilities. In this chapter, we introduce the power and potential of quantum computing.

TIP

This chapter presents many terms and concepts that may be unfamiliar to you. Don't worry; we explain them all in later chapters. (For instance, Chapter 3 describes quantum mechanics and how quantum computing depends on it.) Think of this chapter as boot camp for the new, quantum-computing-savvy you who will emerge after reading this book.

Understanding Why Quantum Computing Is So Strange

Quantum computers have a sense of strangeness about them, almost a mystical aura. (The 2022 movie, *Dr. Strange in the Multiverse of Madness*, captures some of the feeling that people have about quantum mechanics in general.) Why is this?

There are two main reasons. The first reason is people's fundamental misunderstanding of the nature of matter, which quantum mechanics explains. The second is the incredible power that quantum computing, when mature, is expected to deliver to humanity.

How does quantum mechanics (described in Chapter 3) change people's view of the world? The world we live in, where rocks fall down and rockets go up, seems to be dominated by solid matter, with energy as a force that acts on matter at various times. Yet matter can simply be seen as congealed energy.

Most of the mass of the protons and neutrons inside the nucleus of an atom, for instance, is simply a bookkeeper's description of the tremendously powerful energetic fields that keep these particles in place. One of the most important kinds of particles in quantum computing, photons, have no mass at all; they are made up of pure energy.

And it was Einstein himself who told us that matter and energy are equivalent, with his famous equation, $E=mc^2$. To translate: The energy contained in solid matter equals its mass times the speed of light squared.

The speed of light is a very large number — 300,000 km/second, or 186,000 miles/second. Squaring the speed of light yields a far larger number. Plug this very large number into Einstein's famous equation and you'll see that there is a *lot* of energy in even small amounts of matter, as demonstrated by nuclear power plants and nuclear weapons.

The point is that, in quantum mechanics, matter is relatively unimportant; particles act more as bundles of energy. And quantum computing takes advantage of the exotic properties of these particles — ionized atoms, photons, superconducting metals, and other matter that demonstrates quantum mechanical behavior.

The second reason that quantum computers get such a strong emotional reaction is the tremendous power of quantum computing. The best of today's early-stage quantum computers are not much more powerful, if at all, than a mainstream supercomputer. But future quantum computers are expected to deliver tremendous speedups.

Over the next decade or two, we expect quantum computers to become hundreds, thousands, even millions of times faster than today's computers for the problems at which they excel. People can't really predict, nor even imagine, what it's going to be like to have that kind of computing power available for some of the most important challenges facing humanity, as we describe in Chapters 13 and 14. That future is very exciting, yes. But it's also a bit, as Einstein described quantum mechanics, "spooky."

Grasping the Power of Quantum Computing

To help you get started in understanding quantum computing, here are five big ideas to get your head around:

- **Qubits:** *Qubits* are the quantum computing version of bits — the 0s and 1s at the core of classical computing. They have quantum mechanical properties. Qubits are where all the magic happens in quantum computing.

- **Superposition:** While bits are limited to 0 or 1, a qubit can hold an undefined value that is neither 0 nor 1 until the qubit is measured. The capability to hold multiple values at once is called *superposition.*
- **Entanglement:** In classical computing, bits are carefully separated from each other so that the value of one does not affect others. But qubits can be entangled with each other. When changes to one particle cause instantaneous changes to another, and when measuring a value for one particle tells you the corresponding value for another, the particles are *entangled.*
- **Tunneling:** A quantum mechanical particle can instantaneously move from one place to another, even if there's a barrier in between. (Quantum computing uses this capability to bypass barriers to the best possible solution.) This behavior is referred to as *tunneling.*
- **Coherence:** A quantum particle, such as an electron, that is free of outside disturbance is *coherent.* Only coherent particles can exhibit superposition and entanglement.

How are these terms related? Here's an example: A good qubit is relatively easy to place into a state of coherence and maintain in a state of coherence, so it can exhibit superposition and entanglement, and therefore can tunnel. (The search for "good qubits" is the subject of a lot of work and controversy today. We describe this topic in more detail in Chapters 10 and 11.)

These five terms are at the heart of the promise of quantum computing and are involved in many of the challenges that make quantum computing difficult to fully implement. In this section, we describe each of these crucial concepts.

TECHNICAL STUFF

Classical computing describes the computers we use every day, which includes not only laptop and desktop computers but also smartphones, web servers, supercomputers, and many other kinds of devices. The term *classical computing* is used because classical computers use classical mechanics, the cause-and-effect rules of the road that we see and use in our daily lives, for information processing. Quantum computing uses quantum mechanics — which is very different, very interesting, and very powerful indeed — for information processing. We mention some quantum mechanical principles in this chapter and go into more detail in Chapter 2.

Introducing Puff, the magic — qubit?

Bits power classical computing — the laptops, servers, smartphones, and supercomputers that we use today. *Bit* is short for *binary digit,* where *digit* specifies a single numeral and *binary* means the numeral can have only one of two values: 0 or 1 — just like the results of a coin flip.

In a computer, bits are stored in tiny, cheap electromechanical devices that reliably take in, hold, and return either a 0 or a 1 — at least until the power is turned off. Because a single bit doesn't tell you much, bits are packaged into eight-bit bytes, with a single byte able to hold 256 values. (2^8 — all possible combinations of 8 binary digits — equals 256.)

A *qubit* is a complex device that has, at its core, matter in a quantum mechanical state (such as a photon, an atom, or a tiny piece of superconducting metal). The qubit includes a container of some kind, such as a strong magnetic field, that keeps the matter from interacting with its environment.

A qubit is much more complex and much more powerful than a bit. But qubits today are not very reliable, for two reasons:

- They're subject to errors introduced by noise in the environment around them. A result of 0 can be accidentally flipped to a result of 1, or vice versa, and there's no easy way to know that an error has occurred.
- It's hard to keep qubits coherent, that is, capable of superposition, entanglement, and tunneling.

The situation with qubits today is somewhat like the old joke about a bad restaurant: "The food is terrible — and the portions are so small!" With qubits, the error rates are high and the coherence period is short. But despite these problems, quantum computers do deliver valuable and interesting results while up and running.

In quantum computers, qubits are much more complex and far more expensive than bits. Nor are they as easy to manage — but they are far more powerful.

Figure 1-1 shows a quantum computing module from IBM, suspended at the bottom of a cooling infrastructure that keeps the superconducting qubits at a temperature near absolute zero.

Until it's measured, each qubit can represent an infinite range of values between 0 and 1. How does the qubit hold all these values? At the core of the qubit is a *quantum particle* — a tiny piece of reality in the form of a photon, an electron, an ionized atom, or an artificial atom formed using a superconducting metal.

For quantum computing, the quantum particle at the core of the qubit must be kept in a *coherent state* — uncontrolled, like the flipped coin while it's spinning in the air. In a coherent state, we don't know whether the value of the qubit at a given moment is 0 or 1. When we measure the state of the qubit, the calculation we want to make is performed, and the qubit returns 0 or 1 as a result.

FIGURE 1-1: A quantum computing processor from IBM.

Flickr/Lars Plougmann

Much of the power of qubits comes from the fact that they behave in a *probabilistic* manner; a given qubit, running the same calculation multiple times without errors, may produce a 0 on some runs and a 1 on another. The final result consists of the number of times each qubit returns a 0 or a 1. So the result of most quantum calculations is a set of probabilities rather than a single number.

Qubits are hard to create and hard to maintain in a state of coherence; they also tend to interfere with nearby qubits in an uncontrolled fashion. Taming qubits is one of the biggest challenges to overcome in creating useful quantum computers.

A popular approach to building quantum computers involves the use of superconducting qubits, which must be kept at a temperature very close to absolute zero to minimize interference due to heat and, in many cases, to maintain superconductivity.

Classical computers are designed to work at room temperature, but they tend to generate heat and to stop working properly as the temperature rises. The need to dissipate heat prevents device makers from packing components as tightly as they would like without resorting to expensive and clumsy solutions such as water-cooling or refrigerating the components.

In quantum computing, each additional qubit adds exponentially to the power of the computer. But because qubits tend to interfere with each other, adding more is difficult.

IBM, a leader in quantum computing, has published a roadmap showing past and future increases in the number of qubits that power its current and upcoming quantum computers. A simplified version of the roadmap is shown in Figure 1-2. You can find a link to the current version of the roadmap at `https://research.ibm.com/blog/ibm-quantum-roadmap-2025`.

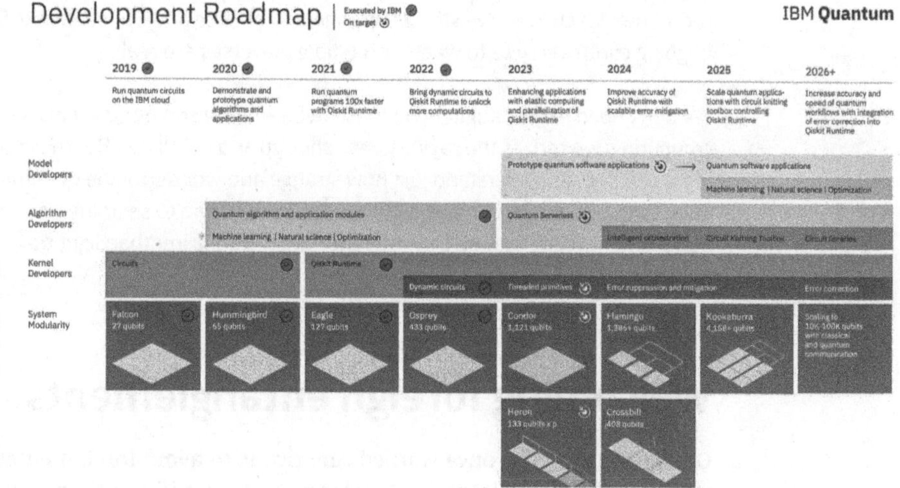

FIGURE 1-2: IBM's quantum computing roadmap shows past and anticipated growth in the number of qubits available.

Superposition, the first quantum superpower

The state of possibility that's available to qubits is called *superposition*, where *super* means *many* and *position* means *possibilities.* A traditional bit can be either 0 or 1. A qubit in a state of superposition does not have a defined value because it holds many potential values at the same time. But when we measure a qubit, we just get 0 or 1 back — whichever value the qubit's energetic wave function collapsed to when it was measured.

Superposition is the first of two major pillars underpinning the power of quantum computing. The other, entanglement, is described in the next section.

In Chapter 3, we describe how the creators of quantum mechanics discovered superposition and other quantum principles during an extraordinary period of scientific creativity between about 1900 and 1930. (This period also saw the disaster of World War I and the beginning of the Great Depression.)

EXPLAINING LIFE, THE MULTIVERSE, AND EVERYTHING

Quantum computing depends on the laws of quantum mechanics, described at a basic level in Chapter 3. Quantum mechanical principles are vital to the operation of quantum computers.

One way of understanding how quantum mechanics works is that each possible state of a quantum mechanical object becomes true — in a new parallel universe that comes into existence to contain that possibility. This way of looking at reality is the basis of recent movies such as *Everything Everywhere All at Once* and *Quantumania*. But there's ongoing controversy as to whether multiple universes are real.

We don't need to get caught up in these debates as we do our daily work in quantum computing; we can, as the saying goes, "shut up and calculate." But these seemingly wild ideas can help us understand just how strange and wonderful the quantum world is. They can inspire us in our daily work, and they may lead to seemingly science-fictional achievements such as time travel, teleportation, and faster than light travel — for information, at least, if not for objects or people.

Welcoming foreign entanglements

George Washington once warned Americans to avoid foreign entanglements. But with qubits, we welcome entanglement as an additional, powerful tool in our quantum computing toolkit.

Entanglement is a kind of connection between two or more quantum particles. For instance, quantum particles have a property called *spin*, which we can measure as either down or up (0 or 1). If two quantum particles are entangled and one of them is measured as having an up spin, we know without measuring that the other entangled particle will have a down spin. And if we influence the spin of the first quantum particle so that it changes to up when it is measured, we know without measuring that the other quantum particle will change to down.

Figure 1-3 shows the connection between two entangled qubits, which have opposing spins. Measuring the spin of one tells you that the spin of the other is the opposite; changing the spin of one qubit in one direction will change the spin of the other in the opposite direction.

As mentioned, entanglement is the second pillar supporting the power of quantum computing. With entangled qubits, influencing a single qubit can have a knock-on effect on many others.

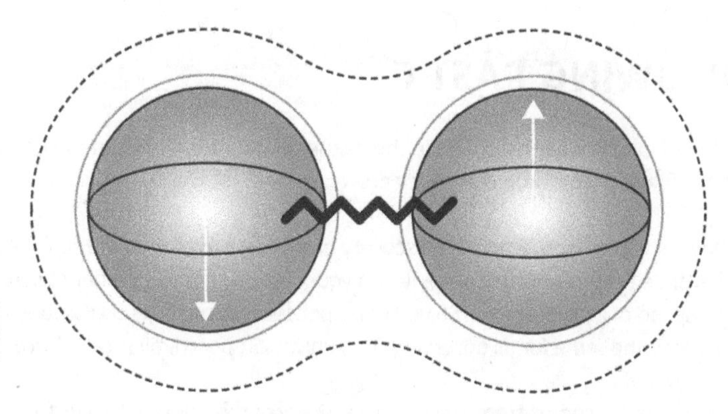

FIGURE 1-3: Entangled qubits influence each other.

Entanglement and superposition work together. When an entangled qubit is in a state of superposition, each of its entangled connections is also in a state of superposition. These cascading uncertainties exponentially increase the potential power of quantum computers.

To program and run calculations on a quantum computer, the potentiality of the entangled qubits must be maintained by keeping them coherent and free from noise. We then measure the qubits (which causes them to decohere) and record the results, a 0 or 1 for each qubit.

TECHNICAL STUFF

Quantum communications is made possible by using entangled qubits for communication between distant locations. We explain this in more depth in Chapter 3.

Enabling quantum computing with coherence

Qubits can be used for quantum computing only when they're kept in a state of *coherence*, free of interaction with their environment. To do quantum computing, qubits need to follow the rules of quantum mechanics (as explained in Chapter 2), and these rules apply to only coherent qubits.

TECHNICAL STUFF

Quantum particles zipping around the universe — photons emitted by the sun, for example — are in a state of coherence. What causes them to decohere? Any interaction with excessive interference (such as vibration or a strong magnetic field), a solid object, or a measuring device.

Keeping qubits coherent is hard. Heat decoheres them, so qubits are kept cold. So do vibration (think of a truck going by on a road) and any collision with their environment. To prevent such collisions, qubits often use strong magnetic fields or targeted laser beams to prevent the quantum particles inside them from colliding with their physical containers.

BLOWING PAST C

Albert Einstein wears two hats in the history of quantum mechanics — and the two hats don't fit comfortably on a single head.

One hat comes from Einstein's discovery of relativity, published in 1905. Relativity says that speed in this universe depends on your motion relative to other observers, but that the speed of light — about 186,000 miles per second, or 300,000 kilometers per second — is always the same for all observers. This universal speed limit is called *locality*.

The other hat comes from Einstein's discovery of the photon, also in 1905. (This discovery, not relativity, is the source of Einstein's sole Nobel Prize.) The discovery of the photon is fundamental to quantum mechanics.

Einstein's problem is that quantum mechanics later asserted that quantum particles, such as photons, can be entangled with each other, so that reading the spin (for example) of one photon tells you the spin of the other. And this relationship is instantly true, without regard to the speed of light. Physicists call this an assertion of nonlocality, which is supposed to be forbidden by relativity.

Einstein hated this, calling it "spooky action at a distance." He and his colleagues spent a great deal of effort trying to disprove it, even as Einstein continued to make breakthrough quantum discoveries, such as the identification of Bose-Einstein condensates, which are superconducting gases that can be used to create qubits.

Today's mainstream computers are subject to classical mechanics and limited by the speed of light. Quantum computers depend on quantum mechanics and, in their use of entanglement, are not limited by light speed.

The Nobel Prize for Physics in 2022 was awarded to physicists who showed that entanglement is real. So researchers in quantum computing who depend on entanglement can say, after Galileo: "And yet it computes." (Galileo, on trial for asserting — correctly, as it turned out — that the earth is not at the center of the universe, is famously said to have whispered: "And yet it moves.")

Decoherence is not the only disaster that can affect qubits. Temperature changes, vibration, or physical interaction may change the value of a qubit in an uncontrolled manner without causing it to decohere. This noise causes errors in the results of quantum computations. Minimizing noise and detecting errors are two of the biggest challenges facing quantum computers.

To manipulate each qubit — to program it, for instance, for quantum computing — the qubit must be controlled in such a way as to adjust its value without causing it to decohere. Magnetic fields and laser beams are among the means used to manipulate qubits without causing decoherence.

When we measure the value of a qubit, two things happen:

- The qubit decoheres, becoming subject to the rules of classical mechanics.
- The qubit's value collapses from somewhere between 0 and 1, inclusive, to either 0 or 1.

The qubit must be reinitialized — returned to coherence — before it can be used again for computing.

Some argue that the potential of quantum computers is very limited — that the level of coherence needed for quantum computers to achieve useful results is impossible, in theory and in fact. In the extreme version of this argument, leaders in quantum computing are accused of deliberately committing fraud, which would mean that the entire field is a massive conspiracy. Only further work will show the limits to quantum computing, if any, but the fraud allegations are just a conspiracy theory.

Doing the Math for the Power of Quantum Computing

It's challenging to fully grasp the potential power of quantum computing compared to classical computing because that power is based on quantum mechanical principles. But we can sum it up in just a bit of math.

Because the bits in classical computing can hold only one of two values — a 0 or a 1 — at the same time, the number of states that a classical computer can hold is represented by the number of bits, n, to the power of two: n^2. But a set of entangled qubits can hold all the possible values of the qubits at the same time. For this reason, the number of states that a quantum computer can hold is represented by two to the power of qubits, n: 2^n. For example, to represent a million possible states would require 1,000 bits but only 20 qubits.

Today's computers contain billions of bits, but we have to throw a lot of them at our most complex problems to get anywhere. Today's quantum computers have a small number of qubits — a recent IBM quantum computer release clocked in with 433 — but we need only a few hundred qubits to begin tackling very complex problems.

The power of today's quantum computers is limited by errors and short coherence times. But as these factors are addressed, the results are likely to be amazing.

Examining What Quantum Computing Will Do for People

It's easy to spend time geeking out on the strangeness and power of quantum computing. But what difference will quantum computing make to humanity?

To understand the answer, we first have to address a common misconception. People today tend to worry about how powerful today's computers *are*: to worry about the power of the internet, social media, and machine learning and AI.

But there's also a big problem around how powerful today's computers *aren't*: They simply aren't up to big computational challenges in areas such as better batteries to fight climate change, better aerodynamics, better routing in complex transportation networks, and better discovery of new drugs, to name a few important examples. (See Chapters 13 and 14 for more information on the challenges that quantum computing will be able to help people to address.)

And these big computational challenges are exactly the areas where we expect quantum computing to make a big difference. Future quantum computers will be able to solve problems we can't touch today, and to do so far faster, more cheaply, and with less energy expenditure than today's computers.

TECHNICAL STUFF

Quantum computers can only "do their thing" in partnership with computers of the kind we use today. So when you see descriptions of what quantum computing can do, understand that these accomplishments will also require a whole lot of conventional computing power.

IS SCHRÖDINGER'S CAT FOR REAL?

No, Schrödinger's cat isn't real. But let's add a bit of nuance to that answer.

Quantum mechanics, the physics that power quantum computing, has been driving people batty since the core theoretical work was done between 1900 and 1930. (See Chapter 3 for details.) And nothing symbolizes this confusion better than Schrodinger's cat.

Erwin Schrödinger, the brilliant physicist behind the Schrödinger's cat thought experiment, was trying to *disprove* quantum computing when he came up with it in 1935. But he actually summed up the mystery behind all things quantum better than anyone, ever.

In the Schrödinger's cat thought experiment, a vial of radioactive poison, some decaying radioactive mass, and a cat are all placed in a box by a scientist. At some point, radioactive decay will release a hammer that breaks the flask, killing the cat, as shown in the figure. But radioactive decay is random, so no one can predict just when the hammer will break the flask.

So the scientist closes the box and goes away, and then waits several hours before coming back. During those hours, an urgent question arises: Is the cat alive or dead?

Quantum mechanics offers only a gnomic answer: yes and no. No intelligence in the world can say for sure whether or not the radioactive decay will kill the cat at a

(continued)

(continued)

particular moment. So the cat is seen as being both dead and alive until the scientist comes back and opens the box.

This seemingly ridiculous conjecture demonstrates *superposition,* in which matter — in this case, the body of a feline — is in two states at once. Superposition, which is critical to quantum computing, has been proven to be real innumerable times since Schrödinger first came up with his thought experiment. (Cats are actually too large, and too embedded within our material world, to be in superposition, in case you were worried. And once again, NO CATS WERE HARMED. Meow.)

A related principle, *entanglement* — a mysterious connection between arbitrarily distant particles, though not cats — has also been proven, in a series of experiments that were recognized with the Nobel Prize for Physics in 2022. It was Einstein who described entanglement in particular, and quantum mechanics in general, as "spooky action at a distance."

So Schrödinger's cat — in the sense of actual cats being placed in mortal danger — is not real. But the principles that animate Schrödinger's immortal thought experiment are very real indeed, as is proven every day in quantum computing.

Describing Different Types of Quantum Computing

There are three ways to do quantum-type computing, and all of them are actively being used and investigated today. They are described in some detail in Chapters 8 through 11; we briefly describe them here.

Quantum-inspired computing

Classical computers are based on logic gates, which take in electrical currents and apply binary logic to them to produce either no outgoing current (a 0) or an easily measurable current (a 1) as a result. By breaking down any mathematical or logical problem to a low enough level and running it through logic gates, you can come up with an answer.

When insights from quantum computing are used to create new algorithms that run on classical computers, or when quantum computer simulators are run on classical computers, this is called *quantum-inspired computing,* as described in Chapter 8. Quantum-inspired computing has proven to be a productive approach to finding new solutions at this early point in the development of quantum computing.

Quantum annealing

In classical computing, there's an approach called simulated annealing. It's analogous to annealing in metallurgy, in which a metal is heated to melt its internal structure, and then cooled to yield a softer, easier-to-work result. *Simulated annealing* is good for solving optimization problems, such as the least expensive route for visiting a number of cities. A similar process to this can be used to solve problems by putting a set of equations through a series of transformations until a result appears.

The quantum computing version of this approach is called quantum annealing. *Quantum annealing* doesn't use a gate-based quantum computing approach, which is more advanced but harder to implement, as described in Chapters 10 and 11. Instead, quantum annealing uses qubits as a group to solve optimization problems, which is a less demanding use of qubits.

Quantum annealers allow for qubit errors and return results that are inexact but still useful in many cases. For instance, a quantum annealer might identify a very good way to route a fleet of delivery trucks, rather than delivering the one and only best possible routing. Only one substantial company, D-Wave, makes quantum computers that use the quantum annealing approach — and they have recently announced plans to make logic-gate quantum computers as well.

Quantum annealers were criticized at one point as not being actual quantum computers, but that is no longer the case. However, gate-based quantum computers get most of the attention, research effort, and investment.

Theory says that quantum annealing should eventually be replaced by a combination of better classical computing approaches — taking over its use cases from "below," in terms of technical sophistication — and gate-based quantum computing, taking over its use cases from "above." We'll see.

Gate-based quantum computing

Most quantum computers use quantum circuitry to reproduce the logic-gate structure of classical computers. These computers are called, sensibly enough, *gate-based quantum computers*. Gate-based quantum computers use qubits in a specific way and are not tolerant of errors in the operation of the qubits; the quantum logic gates don't work reliably if the qubits aren't error-free. Today's qubits are not reliably error-free, which compromises the utility of today's gate-based quantum computers.

Gate-based quantum computing is so hard to achieve that we can't yet be certain that it will ever reach its full potential. But if it does, it will probably render both quantum-inspired computing and quantum annealing obsolete. In the fullness of time, gate-based quantum computing should become the best way to handle any kind of quantum-related computing problem.

Four major types of qubits are used in this type of quantum computer, based on the actual physical object kept in a coherent state to perform quantum computing:

- Superconducting metal loops
- Trapped ions
- Photons
- Neutral atoms

Each kind of qubit has its pluses and minuses, which we describe in Chapters 10 and 11. Different quantum computing companies are basing their efforts on different kinds of qubits.

KEEP YOUR ION THE BIG PICTURE

Atoms are made up of three kinds of particles: protons, neutrons, and electrons. Protons, which are in the nucleus of the atom, have a positive charge of 1. Neutrons, which are found in the nucleus as well and in the same number as protons, have a charge of 0 — no charge at all. The number of protons (and therefore also of neutrons) in the nucleus determines the atomic number of the atom and which element the atom represents. For instance, hydrogen has a single proton and neutron, whereas helium has two of each.

Electrons, which circle the nucleus, carry a negative charge equal in magnitude to the positive charge of a proton: a charge of –1. In most states of matter, the number of electrons balances the number of protons, and the atom as a whole has a neutral electrical charge, because the charges of the protons and electrons offset each other.

In an ionized atom, electrons are added or removed. If an extra electron is captured by the nucleus, the atom now has a negative charge; if an electron escapes the pull of the nucleus, the atom has a positive charge.

Keeping qubits coherent long enough to run logic-gate operations through them is difficult, so logic-gate quantum computers tend to have relatively few qubits — a few hundred at the most, as of this writing — and to run programs for only a fraction of a second before decohering.

Addressing What's Stopping Us

We hope we've given you a taste of the power of quantum computing. What challenges must quantum computing tackle so that all this potential can start delivering more results?

In hardware, several related challenges need to be addressed:

- **Better qubits:** Qubits with lower error rates will produce far better results. Qubits and their control machinery need to generate less noise so as not to cause cross-talk (interference among nearby qubits and the environment) and be better isolated from other noise, such as heat and vibration.
- **More qubits:** Once each qubit behaves better, we can pack them together more densely and in greater numbers. As in classical computing, this development will greatly increase computing power while reducing costs.
- **Cooling and supercooling:** Superconducting qubits require extremely low temperatures, but lower temperatures are better for error reduction for all kinds of qubits. Therefore, better, cheaper, easier-to-manufacture cooling and supercooling will help the other hardware challenges.
- **Error correction:** Error correction will require multiple solutions, one of which is to combine several physical qubits into a single, self-correcting logical qubit. Colder temperatures and better control machinery will reduce the error count and ease the burden on error correction.

In software, major challenges include the following:

- **Quantum algorithms:** Many decades of work have gone into optimizing algorithms for mainstream computers, but quantum algorithms are in their early days. Both theoretical advances and optimization in practice are needed.
- **Hybrid quantum classical computing:** Advances in both software and hardware are needed to help quantum and classical computers share problems smoothly and with high performance; this kind of collaboration between systems is called *hybrid quantum classical computing.*

» **Error mitigation:** Software can help predict and reduce errors, making today's quantum computers more useful and reducing the burden on hardware error correction.

People power is lacking as well:

» **Scientists and researchers:** Theoretical work and advanced development are needed to move the field forward, but there are too few qualified people.

» **Software developers:** Only a small number of software developers know how to use quantum computing systems and contribute to their further development, though this share is slowly increasing.

» **Business analysts and end users:** The user community doesn't yet know what problems to try on quantum computing systems and how to ask for needed improvements that might deliver value in the near term.

Finally, quantum computing is a new way of looking at problems — and often this new viewpoint leads to better solutions for use on classical computers. Only continued work will give everyone involved a better idea as to how to use each kind of computing — and hybrid systems that use both kinds of computing — to the best advantage.

All of this may leave you wondering just where quantum computing stands today. It's actually at an awkward place: showing tremendous potential and attracting billions of dollars in investment worldwide but not yet provably superior to classical computing on important problems.

However, as this book is going to press, IBM has announced work that demonstrates what IBM calls *quantum utility*, which is the capability of quantum computers to achieve useful results that aren't possible on the computers we depend on today.

In this work, IBM used one of their more recent quantum computers, boasting 127 qubits, to simulate the interactions among 127 bar magnets — a subtle, fractal nightmare of a problem for classical computers. IBM ran the same problem across classical and quantum computer systems. Supercomputers (powerful classical computers) were able to keep up with IBM's quantum computer to about 63 qubits, half its capacity. Beyond that, the quantum computer's results seem to have exceed what's possible with supercomputers.

There's still work to do to confirm or deny this apparent breakthrough — and it's only immediately useful to anyone whose product roadmap has exactly 127 interacting bar magnets in it. But as an earnest of intent — that is, "a sort of down payment . . . a promise that you will follow through, and evidence that you can follow through," as one website defines the term — it's pretty darn awesome.

» Tracking the rapid rise of today's computers

» Combining classical and quantum computing

Chapter 2

Looking Back to Early and Classical Computing

Why talk about classical computers — the laptop and desktop computers, tablets and smartphones, web servers, and even the world's largest supercomputers, all of which people depend on every day — in a book about quantum computing? Three reasons, really.

The first is that quantum computers handle only certain specialized tasks, which means they'll never fully replace classical computers. You'll use classical computers to set up and initiate quantum computing jobs, and you'll receive the results via classical computers. So using quantum computers requires using classical computers.

Second, the best way to understand quantum computing is by comparing it to classical computing. So having a firm grasp on the history and progress of classical computing will help you to understand what's so special about quantum computing.

Finally, many of the most challenging tasks for quantum computers will require a hybrid operation, with classical computers and quantum computers handling specific steps in turn. Only by knowing what each kind of computer is best at will you be able to understand these hybrid jobs, which may turn out to be some of the more interesting projects that quantum computing enables.

New supercomputers are already being built and delivered with very fast interconnections to enable a kind of Vulcan mind meld with quantum computers, enabling very efficient joint processing of tasks.

And, if we may add a PS, it's also just fun to review the history of classical computing. It's such an important part of all of our lives, with an ever-increasing daily effect, that having a grasp of the overall story is rewarding. We certainly enjoyed writing this chapter.

Understanding Why Classical Computers Are Not Going Away

Before we dive in, let's try for a definition of what a computer is. The word has actually evolved over the years.

Broadly speaking, the word *computer* has been used to describe anything — or anyone — performing computational tasks. A person who performed computational tasks could be a computer. The abacus, slide rule, or desktop calculator (if you're not old enough to remember them, don't ask) that a computer-who-was-a-person used could also be called a computer.

But when speaking about computers today, we're talking about dedicated devices that focus on performing computational tasks. A computer that performs computational tasks using classical — that is, non-quantum — mechanics is called a classical computer.

Classical computers run today's world. You find them all around you — not only in the more obvious computers just named but also in talking children's toys, robotic vacuum cleaners, and electric vehicles (EVs).

Actually, EVs contain multiple classical computers, working together, but you get the point.

Why do we call these everyday devices classical computers? A *classical computer* is a computer based on the deterministic principles of classical mechanics, as described in the next chapter. You give it data, fire up a program (or recalculate a spreadsheet), and it produces the same results. Every. Single. Time.

Classical computers are inexpensive, reliable, maintainable (or replaceable), and highly capable for most problems. However, with several types of problems, classical computers hit a wall — but in these same circumstances, quantum computers are likely to do very well indeed.

So the best way to lay the groundwork for understanding quantum computing throughout the rest of this book is to take time in this chapter to review the early history of computing and the emergence of today's classical computers.

TIP

There's a lot of controversy about the future value of quantum computing, but it reminds us of the story of an optimistic child who was shown a large pile of horse manure and asked to say something good about it. Instead, the child went and got a shovel and started digging. Puzzled, people asked the child why. The child said, "With all this horse manure, there must be a pony in here somewhere!" Similarly, we believe that all the time, attention, and money being invested in quantum computing are likely to result in significant value to humankind in the not too distant future — even if some of it does turn out to be horse manure.

Looking Back to the Prehistory of Computers

People have been inventing and using computing devices for a long time. We can look at three early computer devices as examples of how thoroughgoing these efforts have been:

- **Abacus:** Beads on wires in a wooden frame, used for arithmetic calculations. Invented thousands of years ago and once used across much of the world; still in some use today.
- **Antikythera:** A shockingly advanced astronomical simulator built by the Greeks about two thousand years ago. Forgotten by history until a single example was discovered in an ancient shipwreck in 1901.
- **Slide rule:** A device with three geared rulers, used for basic and logarithmic calculations. Invented a few hundred years ago and widely used until recently. Still in use by a few enthusiasts today.

Counting on the abacus, a forerunner of the classical computer

The best-known and most widely used ancient computing device is a kind of counting table — not unlike a backgammon board — invented thousands of years ago in several cultures. These counting tables evolved over time into the abacus.

An *abacus* is a series of beads on wires, as shown in Figure 2-1. The user moves the beads to record numbers and to do calculations. The abacus is still used by some shopkeepers and as a toy and teaching device for children, as you'll see if you wander around a toy store or a teacher's supply store.

FIGURE 2-1: The abacus is amazingly fast and accurate in the hands of an experienced user.

Alexey Kuznetsov / Adobe Stock

Later versions of the abacus use a place system and work in a way that's recognizable to us today, although experienced abacus users use shortcuts and tricks that befuddle anyone less experienced.

The abacus has many similarities to today's computers. It's compact, exact, and deterministic. (That is, every time you start with the same inputs to an abacus operation and follow the same steps, you get the same results.) Those beads on a wire do not exhibit superposition, which we introduce in Chapter 1. Entanglement? Nope. Tunneling? Not so much.

TECHNICAL STUFF

The abacus can also be used for approximate operatins and to get to a "good enough" result when large numbers are involved. But it's most often used to get exact answers.

The abacus is still in use today, largely by people who grew up with it. For most purposes and for most people, though, it's been replaced by pocket calculators and calculator apps on a mobile phone.

FINDING OUR PLACE IN A PLACE SYSTEM

A base 10 place system — the decimal digits we use today, and the place system we use to organize them — are ingrained in us. But the truth is that worldwide standardization on the base 10 place system is a relatively recent invention.

The Romans used an arcane system in which the numbers one through five were written I, II, III, IV, and V. Each I represents one and each V represents five. But for four, the Romans would write IV, meaning one less than five. The Romans did not use a place system and did not have the concept of zero as a number. (If you tried to call a Roman a zero — even in Latin — they would just look at you in puzzlement.)

For many people today, Roman numerals appear in their lives only because the Super Bowl, the finals of American football, uses Roman numerals to convey a sense of majesty around the big game. And schools sometimes teach Roman numerals to make the point that our current place-based decimal system is not the only possible way to write numbers.

The fiftieth Super Bowl, which took place in 2016, was widely advertised by its Roman numeral designation, Super Bowl L. (So non-football fans, or fans of the losing Carolina Panthers, could say, "The L with the Super Bowl.")

Simply describing Roman numerals is enough to show key advantages of the decimal digits 0 through 9. But the place system is what gives decimal digits their power.

For ten, we write 10. We know that the 1 in 10 means ten because another digit, 0, is after the 1. The 1 in 100 means one hundred because two digits, both 0s, are after the 1, and so on. (You can see why 0 is very much needed as a placeholder in a place-based system.)

Our decimal place system is the offspring of place systems used by the Babylonians, Mayans, Chinese, and Indians, which varied in which number they used as a base. Today, in classical computing, binary uses base 2 (one bit), octal uses base 8 (three bits), and hexadecimal — so important in the 2015 hit movie *The Martian* — uses base 16 (four bits, sometimes called a nibble because it's half a byte).

Unearthing the Antikythera, an early orrery

There's a long history of people building computational devices to simulate the movement of the heavenly bodies. (Devices of this type, called simulators, are different from calculating devices such as the abacus and the slide rule. You'll see this distinction come up in quantum computing too.) Astronomical movements were important to people both for practical reasons, such as knowing when to plant and harvest crops, and for use in astrological predictions.

These devices are called orreries (pronounced "awr-uh-rees"), and many of them were complex. Before Galileo, they had to be complex because they were based on an incorrect assumption: that the Earth was at the center of our solar system. This incorrect assumption made it fantastically difficult to build a model (centered around the Earth) that matches reality (in which planets circle the Sun).

Orreries have many of the same requirements as clocks, such as the need for a power source and precise gearing. So historians have believed that orreries and clocks evolved together, with slow advances to both through the Dark Ages and the Middle Ages (very roughly, about 500AD to 1400AD), and rapid progress during the Enlightenment that followed. (The Enlightenment is also known as the Age of Reason.)

Yet a 2000-year-old device discovered in a Mediterranean shipwreck in 1901 off the Greek island of Antikythera has upended these assumptions. The device and the shipwreck both date from one or two centuries before the birth of Christ. And the corroded, dented device, which historians and scientists are still examining today, seems to have been a fantastically complex orrery.

Nothing in the archaeological nor written record had led anyone to believe that the Greeks and Romans had a device this complex, let alone one that was a kind of computer. Because the Antikythera (shown in Figure 2-2), like all orreries, is a mechanical simulator — a type of computing device that people create to model a natural process.

The Antikythera had dozens of gears and seems to have been designed to predict solar and lunar eclipses, among other functions. Nothing as complex appeared again until advances in clockmaking more than a thousand years later, in the 1300s. The complexity of the device suggests that there were predecessor and companion devices to the Antikythera, but again, we don't have any written or archaeological proof.

Today, as with all orreries, the Antikythera has been largely replaced by a raft of tools that provide the same information but more conveniently: printed almanacs, online documents, "Googling it," or asking ChatGPT for an answer.

However, it's worth noting that Richard Feynman's initial suggestion for quantum computers — made in 1959 and repeated in 1981 — was for their use as simulation devices: to model quantum mechanical interactions as found in drug discovery and materials science, for example. Today, classical computers, graphical processing units (GPUs), specialized quantum devices, and fully capable quantum computers are used both for calculation and as quantum simulators.

Marsyas / Wikimedia Commons / CC BY-SA 3.0

FIGURE 2-2: The Antikythera is still slowly yielding its secrets.

Calculating why the slide rule no longer rules

The slide rule is a more recent example of a computing device, though to many of you reading this, it might also seem like ancient history. Shown in Figure 2-3, the slide rule is a set of rulers with markings used for distances, to represent numbers, and for logarithms (exponential powers of a number). As with an abacus, a slide rule comes to life only in the hands of a skilled user.

Slide rules were invented in the 1600s but became commonplace among engineers in the mid-1800s. The movie *Hidden Figures* (2016), based on historical fact, features black American women who worked as human computers, using slide rules for the orbital and other calculations needed to get astronauts into space and to the Moon.

Buzz Aldrin is said to have whipped his out, while in lunar orbit, to help with last-minute calculations that brought the Eagle safely to rest on the surface of the Moon. *The Right Stuff* — a 1983 movie about the Space Race — indeed!

As with the abacus, slide rules were eventually replaced by pocket calculators in the 1970s, and are now largely supplanted by mobile phone apps, online searches, and computers. Also like the abacus, the slide rule still has a core of users, most of whom grew up with the device. A few younger people use them as well but rarely with the deftness of their elders.

FIGURE 2-3: Slide rules once ruled, from artillery ranges to Cape Kennedy to lunar orbit.

dvande/Adobe Stock

The slide rule is exact for small numbers and approximate for most large-number calculations. (A quick approximation often had to be good enough because it would take hours or even days of work with pencil and paper, or the use of precious mainframe computer time, once those were invented, to produce a more exact result.) This useful inexactitude of many slide rule calculations is somewhat analogous to today's quantum computers, which often give inexact yet highly useful results.

Assessing what we can learn from early computers

This brief survey of selected early computers shows that a lot of different kinds of computing devices are, or have been, in existence. And a given computing device, once it does come into use, may be capable of results and evolution that the original builders could never have imagined. (For example, the slide rule came into use well before 1865, when Jules Verne popularized the idea of people going to the moon; but the slide rule, little changed, ended up helping to make the moon landing possible a century later.)

Similarly, we see an interesting dynamic between classical computers, described in the next section, and quantum computers. The development of quantum computers, and quantum computing algorithms to run on them, has, in turn, inspired a lot of recent progress in classical computing.

Beyond these specifics, the main purpose of understanding the history of computing and the workings of current classical computers is to inspire us to think broadly about what we can do with computers of all kinds and quantum computers in particular. Together and separately, classical and quantum computers are certain to further change our world, but it will take human effort and imagination to power those changes.

Tracking the Emergence of Classical Computing

Classical computing uses the principles of classical mechanics — the deterministic approach used by Copernicus, Galileo, and Newton, as described in Chapter 3. Today, after more than a century of development, mainstream classical computers are capable, cheap, and reliable. They're also challenged by certain kinds of problems, which we describe later in this chapter.

The history of computing can be described in many ways, but we focus here on a few key advances that describe how we got to the world we live in and set the stage for our main interest, the emergence of quantum computing. Those milestones are

- The arrival of electromechanical tabulating machines
- The first electronic computers
- The rise (and rise) of the microprocessor

Counting on the arrival of tabulating machines

The direct ancestors of today's computers were tabulating machines, developed to help total the US Census in the late 1800s. (Tabulating machines, though limited, do arithmetic, which is a basic information-processing task, as do the abacus and the slide rule.) The late 1800s was the era of the railroad, the first metal ships, and — just after the turn of the century — the first airplane.

This is also the world in which quantum mechanics was developed, between about 1900 and 1930, as described in Chapter 3. The inventors were wearing the clothes and using the technology of the Victorian era while creating mathematics and physics that even people today don't fully understand. It's ironic that many of today's quantum computers, encased as they are in chandelier-like supercooling machinery, have such a steampunk look.

Steampunk is a literary genre in which science fiction stories are set in the era of steam-powered machinery, resulting in complex and impressive — but imaginary — machines far more capable than those which people of the time possessed. Think of the machinery used to bring Frankenstein's monster to life in various movies, with electricity arcing through the air and clouds of steam being thrown up, and you'll have the idea. Today, steampunk is widely used to describe quantum computing to newcomers because of those chandeliers and other impedimenta — and because it feels like we're living in an old and a new era at the same time; there's even a book with the title *Quantum Steampunk*.

Now, back to the development of classical computers. A few key points about tabulating machines:

- They were the offspring of mechanical looms (devices for weaving cloth) and player pianos, both of which used holes punched in rolls of paper as programming.
- They took dollar-bill-sized punched cards as input (punched cards continued in widespread use long enough for one of the authors, Smith, to have started his career by tabulating them with a timesharing computer in the 1980s).
- Their initial purpose was addition, to arrive at totals for various fixed fields.
- They were electromechanical: powered by electricity but mechanical in nature, using pulleys, gears, and so on to do their work (shades of the Antikytheron, described previously).
- They were the main form of computing until the electronic computer arrived in World War II, and faded only gradually after that.

Tabulating machines were born in a way similar to today's Silicon Valley-style startups. By the late 1800s, tabulating the results of a census using pencil and paper was expensive, error-prone, and time-consuming, taking nearly a decade. Since the census is taken every ten years, this was not, as we would say today, a good look.

So a census employee, Herman Hollerith, quit government work and started a tabulating machine company. The government contracted with him, and the 1890 became census was the first to be tabulated by machine. Hollerith became very successful.

The Hollerith Tabulating Machine Company went through a series of mergers and acquisitions until it became part of the Computing-Tabulating-Recording Company, founded in 1911. The company was renamed International Business Machines (IBM) in 1924.

Tabulating machines continued to be used by the census for decades; Figure 2-4 shows one in use to tabulate results of the 1950 census.

FIGURE 2-4: A tabulating machine and its operator add results from the 1950 census.

U.S. Census Bureau / Wikimedia Commons / Public Domain

It's hard to imagine today how dominant IBM was in technology for most of the 20th century. And not only for their products — IBM's management style, their semiformal dress code (featuring blue suits, white shirts, and ties), and their corporate slogan, "Think," were famous across the US and around the world.

Examining a mathematical model for classical computers

A great deal of mathematical theory underpins the creation and use of computers, including classical computers. We describe only one crucial part of that theory here: the invention of a theoretical model called the Turing machine. Alan Turing, who created the Turing machine, was also the central figure in the development of the first electronic computer.

The computers that Turing developed used electromechanical relays and vacuum tubes, since transistors were not invented until 1947. The use of relays and vacuum tubes made the computers unreliable, but it was the flight of a moth into the circuits of an early computer, bringing its operations to a halt, that led to the term *computer bug*.

Today's classical computers are real-world, limited implementations of a Turing machine, which was invented by Turing in 1936 for his doctoral thesis at Cambridge. This mathematical model can perform any imaginable set of logical or mathematical operations by having them broken down into simple steps that the machine then carries out.

REVEALING THE STRANGE FATE OF IBM'S COMPUTER BUSINESS

To jump ahead in our story by a few decades, IBM caught the wave through three major technology transitions, across nearly a century. But they wiped out on the fourth one, 100 years after Hollerith's initial success.

IBM grew powerful with the rise of tabulating machines in the early 1900s. And they successfully made the move to mainframe electronic computers after World War II. Then they elbowed their way to leadership in the minicomputer business after upstarts such as DEC and Univac challenged IBM in the 1950s and 1960s.

IBM then sought to achieve leadership in another new architecture, the microcomputer, or personal computer, with the IBM PC in the early 1980s. But the company lost control of the market.

The initial IBM PC depended heavily on the familiar IBM name and salesforce to get traction. But the company counted on two key suppliers: Intel for microprocessors and Microsoft for the operating system (PC-DOS). Intel and Microsoft then went around IBM and worked with independent computer vendors such as Compaq to commoditize the architecture and compete with IBM.

As a result, IBM lost its leadership position in PCs, as Compaq seized first place in 1994. Eventually the entire PC line of business lost money. In 2005, IBM sold its PC business to Lenovo, which for a while became the leading PC company in the world.

Today IBM sells a wide range of hardware and software products as well as consulting and business services. The company is a leader in open-source software, data analytics, supercomputers, and AI, among other areas. The company's efforts in quantum computing may move IBM back to the forefront of computing.

A Turing machine is an ideal machine, rather than an actual, physical one. This ideal machine has a read-write head that is always in some state among a finite number of states. The read-write head is positioned over one cell in an imaginary tape that is an infinite number of cells wide and long. (We told you it's a mathematical model; in our limited and imperfect world, an actual infinite tape is impossible.) Each cell in the tape holds one symbol from a finite set of symbols.

Each step begins with the read-write head positioned over a single cell of the tape. The read-write head reads the symbol in the cell below it. Depending on its current state, the read-write head then writes a symbol back into the cell.

The contents of the cell also tell the read-write head to move one step left, move one step right, or stay in the same place. And the contents of the cell may tell the tape to change its state, so it will begin to carry out a different set of operations in response to the next cell's input. The tape then advances one step, putting a new cell below the read-write head. By following these simple instructions, this ideal Turing machine can, given enough time, complete any logical or mathematical operation that the human mind can conceive of.

Work on the Turing machine created a theoretical framework that we continue to depend on today in both classical and quantum computing. But the first practical use of the Turing machine framework was in cracking codes produced by Enigma, an enciphering machine used by the German armed forces. An Enigma machine is shown in Figure 2-5.

FIGURE 2-5: A World War II German Enigma cryptographic machine.

It's hard to overstate the importance of having a mathematical model for something like the operation of a classical computer. With the model available, there's always a theoretical reference point for any relevant machine you want to build. A group can spend billions of dollars, secure in the knowledge that what they're trying to achieve is theoretically possible; the only remaining problems are in the realm of engineering. (The development of the electronic computer during World War II implemented much of Turing's theoretical work, and the development of the atomic bomb implemented part of Einstein's.) Without a model, everything must proceed by trial and, all too frequently, error.

ARE TURING MACHINES PASSING THE TURING TEST?

You may have heard the phrase *the Turing test,* which was popularized in the movie *The Imitation Game* (2014). The phrase is related to the Turing machine, but it was coined slightly after the Turing machine was defined in the 1930s.

As the Turing machine concept was first being used as a reference architecture for electronic computers, Turing's colleagues began to wonder if a computer could someday become as intelligent as a human being. They also wondered how a computer could prove it had reached that benchmark.

They weren't wondering if a computer could beat people at specific tasks; even Hollerith's machines in the 1890s could beat people at calculating, for instance. They were wondering if a computer would someday be able to display that certain *je ne sais quois* that distinguishes our interactions with humans from our interactions with machines. This high level of capability is crucial to what AI researchers and others today call artificial general intelligence (AGI).

So Turing proposed the Turing test: A tester would sit at a teletype and carry on teletyped conversations with a human being and with a computer. If the computer's answers were as intelligent as the person's, and the tester could not tell which was which, the computer would have passed the Turing test.

For decades, the Turing test produced results that were visibly odd, which we humans found reassuring. Even when a computer beat the best human grandmaster in chess (1997) or won on the TV game show *Jeopardy!* (2011), no computer could repeatedly pass the Turing test.

However, the emergence of ChatGPT in 2023 shows computers on the verge of passing the Turing test. (ChatGPT has, for instance, managed to pass the American bar exam, required for admission to law school.) It may not be long before the visible difference between a computer and a person in conversation will be no difference at all.

Classical computers are Turing machines, with two exceptions: They use RAM instead of tape — though RAM can be used in a way that emulates tape — but the amount of RAM or tape available is not infinite. Yet most computer programming languages are "Turing complete," which means they are capable of carrying out any imaginable set of logical or mathematical operations. Since the machine the programs run on is definitely not Turing complete, due to its finite memory, many of the most devilish programming bugs are due to running out of memory while running an otherwise valid program. (Later computers advanced to the point where they could also fail by running out of disk space, and later ones still due to network failures, all of which are different versions of the same problem.)

The Turing machine just described is deterministic, as are classical computers; for instance, a classical computer can't generate a true random number. This determinism is a huge advantage for many tasks; "store the contents of the Library of Congress and retrieve specific information for me when I request it" is a good thing to be deterministic about. But determinism is a severe limitation for other tasks, which is where quantum computers come in, as we describe in Chapter 6.

Commemorating the first electronic computer

Today's electronic computers use electronic components, rather than mechanical relays or vacuum tubes, to perform computing tasks. The first electronic computer, Colossus, was created by the British — with Alan Turing playing a leading role — to crack German codes in World War II.

Tragically, after World War II, Turing was persecuted for being gay. He was forced to take hormone injections that left him in a state of depression and led to his suicide in 1954. (Another British genius, Oscar Wilde, was persecuted to death for the same reason half a century earlier.)

Colossus and its immediate successors performed computing tasks using vacuum tubes, which were used at the time to power a wide range of devices, including televisions. But between roughly the 1940s and the 1960s, vacuum tubes were largely replaced by transistors.

TECHNICAL STUFF

Ironically, both vacuum tubes and transistors are among the many advances made possible by quantum mechanics. Both use quantum mechanical principles to carry out deterministic tasks. In fact, making sure that transistors do not display quantum mechanical properties, such as allowing electrons to tunnel from one logic gate to another unpredictably, is a key challenge in the design and manufacture of classical computers. (Tunneling is increasingly hard to avoid at very small sizes, so its emergence as sizes decrease is preventing classical electronic circuits from getting much smaller.)

During the transistor era, the basic architecture of a classical computer emerged; elements of this architecture are shown in Figure 2-6. Many variations exist, but the architecture usually consists of the following components:

- **Central processing unit (CPU):** The CPU uses logic gates to carry out instructions. The logic gates are chosen in such a way that any imaginable series of steps, whether mathematical (add all these numbers) or logical (sort these names in alphabetical order), can be carried out by the CPU.
- **Random access memory (RAM):** RAM, which is fast and relatively expensive memory, stores the most frequently accessed program code and data that the computer is currently using. RAM typically empties out when the power is turned off or the machine is shut down (though more and more devices use static RAM, which does not clear out).
- **Hard disk drive:** A hard disk drive is a spinning magnetic disk that holds static copies of program code and data. It's roughly 100 times slower than RAM but also roughly 1,000 times cheaper. Importantly, the hard disk drive doesn't lose its contents when the power goes out. Many disk drives are being partly or completely replaced by flash RAM, which is smaller, faster, and uses less power.
- **Data buses:** Data buses are not wheeled vehicles that carry information. They are, rather, predesigned hardware pathways for the rapid movement of data (including program code) within the computer.
- **I/O devices:** Input/output devices such as a keyboard, touchpad, and display screen allow the user to interact with the computer.
- **Peripherals:** Peripherals are external devices, such as printers or smartphones, that plug into the computer via a connector such as a USB port or a Bluetooth wireless connection.

This basic architecture has been used for computers from the mainframe, introduced in the 1950s, to the microcomputer, introduced in the 1980s. And a version of this architecture is used in the tablets and smartphones of today. But one particular component has done more than any other to make classical computing unavoidable in our daily lives: the microprocessor, which we describe right after we introduce the bit and the byte.

Understanding the invention of byte-sized data

Classical computers famously use a *bit* — a single unit of information that can be set to 0 (off) or 1 (on). But classical computers tend to work with groups of eight bits, called a *byte*. Why is this?

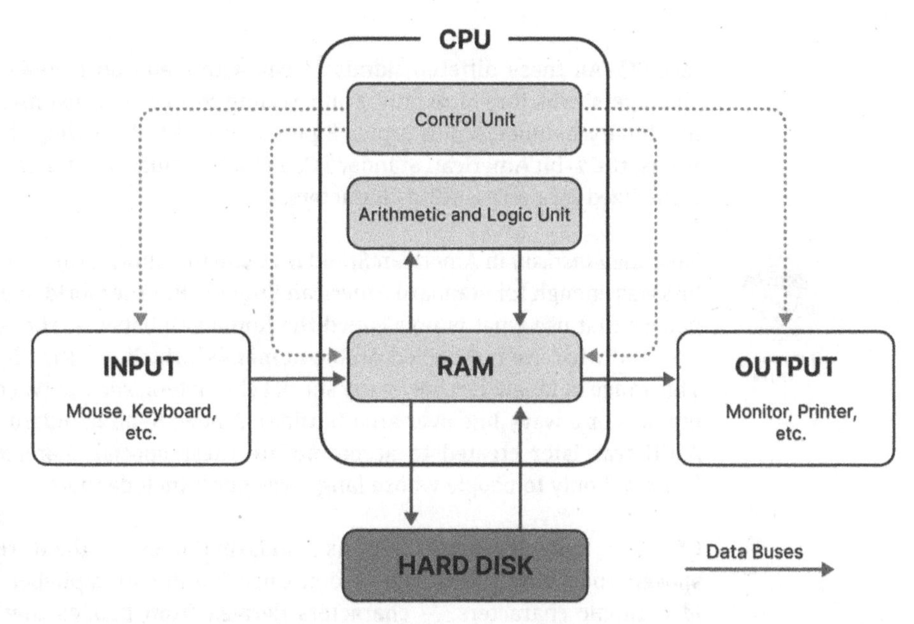

FIGURE 2-6: Architecture of a typical classical computer.

First, it's important to understand that binary numbers use a place-based system, just like the decimal numbers we use every day. However, the places mean something different.

In decimal numbers — numbers that use ten digits, 0 through 9 — every new place is an additional power of 10. In the decimal number 837.4, 8 represents 800 (because the 8 is in the third place), 3 represents 30, 7 represents 7, and .4 represents four-tenths of 1. If you rearranged the digits, you would have a much different number because of the power of place values in a number.

In binary numbers, you have only two digits, 0 and 1, so each additional place represents an additional power of 2, not of 10. In the binary number 1111, the first 1 represents 8, the second 1 represents 4, the third 1 represents 2, and the last 1 represents 1, for a total (in decimal numbers) of 15.

So why use 8-bit bytes? Because bytes are a convenient size for representing a fairly wide range of alphanumeric characters and can easily be combined to represent a wide range of numbers.

If you look at this page one character at a time, your eyes will quickly glaze over. However, if you persevere, you'll see most of the letters of the Roman alphabet in lowercase (*a* through *z*), many of them in uppercase (*A* through *Z*), many digits (0 through 9), and a few different punctuation marks (,.-!) and special characters

(@$^*). All these different kinds of characters add up to fewer than 128 total characters, which require only 7 bits. (Going from the lowest place to the highest in a binary number, 2 plus 4 plus 8 plus 16 plus 32 plus 64 equals 128). So, in the 1960s, the 7-bit American Standard Code for Information Interchange (ASCII) was formalized for representing characters.

Note the *American* in American Standard Code for Information Interchange. Seven bits was enough for standard American English. But the world includes many languages that use what is now called the Roman alphabet — the letters *a* through *z* — and what are now called Arabic numerals, which use the digits 0 through 9. This range of languages has many special characters, such as two dots over a *u* (an umlaut) or a wavy line over an *n* (a tilde). A new, 8-bit standard called Extended ASCII was later created to accommodate these special characters. (Which are "special" only to people whose languages don't include them.)

Of course, many languages, such as Mandarin Chinese — the world's most widely spoken and written language — don't use the Roman alphabet at all. They use ideographic characters — characters derived from images that once conveyed ideas but are now largely used in an abstract way. (People who speak, read, and write Chinese can come up with some fairly incredible puns by using the dual nature of these characters.)

To accommodate all the characters used around the world, a new and extensible standard, called *Unicode,* was created. Unicode uses 16 bits, allowing for up to 65,536 characters (that's 2^{16} as a decimal number). That's enough characters for now — but if we someday communicate with nonhuman intelligences, whether they be extraterrestrial, canine, feline, delphine, or even silicon-based, Unicode may need to be extended further.

Around the time that ASCII was being defined, IBM introduced an 8-bit code for card punches used with its System/360 line of computers. The 7-bit ASCII code and the 8-bit card punch code were close enough in bit count that 8 bits could work for both purposes, and an 8-bit byte became the standard.

Numbers are a whole different beast. There are many different formats for storing numbers, depending on the needs of a particular computer program. The standard for floating-point numbers was set at 32 bits, or 4 bytes. A standard floating-point number can exactly store any value between about negative 4 million and positive 4 million, and can store approximations of most larger positive or negative numbers.

Binary digits are also very important in quantum computing, as you'll see in Chapter 4.

Tracking the rise of ICs and microprocessors

Early computers used individual transistors, artfully arranged for maximum performance. Around 1960, the workhorse of the digital era, the integrated circuit (IC), was developed. ICs combine a number of transistors on a single chip.

The IC gave rise to the most important component of today's classical computers, the microprocessor. Microprocessors were pioneered in the 1970s and initially used to power electronic calculators. As microprocessors became more powerful, they were used for the first personal computers, including the Apple I, the Apple II, the IBM PC, and the Macintosh, all between about 1975 and 1985.

Figure 2-7 shows an Intel 80486, an important early microprocessor. The 80486 was the first microprocessor with more than one million transistors and had a feature size — the distance between one end and the other of a single transistor — of less than a micron (0.8 microns), both impressive milestones for the time. The Intel 80486 arrived in 1989 and was the first microprocessor that could deliver performance good enough to run the then-current version of Microsoft Windows, which is still in use today.

FIGURE 2-7: The Intel 80486, an important microprocessor.

The microprocessor is the most visible beneficiary of Moore's Law, which states that the power of computer chips of a given size and cost doubles every two years or so. This so-called law — which was initially an observation, and then became, for many decades, an accurate prediction — propelled computing from a small part of the economy before World War II to become perhaps the greatest single driver of economic growth around the world.

The microprocessor is also crucial to the single most important new development in computing, the range of technologies called artificial intelligence (AI) and machine learning (ML). Specialized processors called graphics processing units (GPUs), originally popularized for computer games, often handle the grunt work of AI and ML at high speed.

But even GPUs are overtaxed by the ambitions of today's AI and ML programs. AI and ML are likely to be among the main beneficiaries of quantum computing as it matures.

Joining Classical Computing and Quantum Computing

We describe quantum computing's development in Chapters 5 and 6. But we can make some valuable comparisons here:

- Quantum computers are used for a task only if classical computers can't perform that task in a reasonable amount of time or at a reasonable cost. As a result, most people who interact with quantum computers do so through classical computers.
- Unlike with classical computers, relatively few companies own quantum computers. Most access quantum computers through the cloud on a pay-as-you-go basis known as QaaS (quantum as a service). You can use a cloud portal of a specific quantum computer manufacturer or a cloud service such as Amazon's AWS Braket, Microsoft's Azure Quantum, or the Strangeworks portal.
- Quantum computers are made up of qubits and devices for manipulating them, which serve as the equivalent of a classical computer's CPU and (to some degree) memory.
- Qubits and classical computing CPUs often share tasks. Classical computing technology such as RAM, hard disk drives, and data buses are used to support interaction between a quantum computer, with its qubits, and a classical computer, powered by one or more microprocessors.

IN THIS CHAPTER

» Embracing uncertainty

» Making history, coherently

Chapter 3
Examining the Roots of Quantum Computing

Quantum mechanics is not only a scientific accomplishment of the first order but also a philosophical revolution, changing how we see the universe — in ways that people are still disagreeing about today. Our assumptions about the way the world works are, quite naturally, based on our daily experience and traditional scientific methods developed over centuries. But quantum mechanics challenges those assumptions.

At the core of scientific practice is repeatability: If you do the same experiment with the same starting conditions many times, you will get the same result every time. The same is true in daily life: If you hold a heavy weight above your foot and let it go, gravity will always cause it to fall, and the impact will always hurt. Trust in this reliability is called *determinism*.

But quantum mechanics tells us that the fundamental state of the universe is non-deterministic. When you run experiments on the smallest particles, you can state the probability of different outcomes, but the specific results will vary from one run of the experiment to the next. With quantum mechanics, you can predict the *probability* of each of the possible outcomes, often to many decimal places of precision, but the specific results will rarely come out the same twice.

Quantum computing uses some of the most challenging and controversial assumptions of quantum mechanics to do computing in an entirely new way. As you enter into the new world of quantum computing, understanding the principles of quantum mechanics will help you focus on the right kinds of problems, solve problems more creatively, work around roadblocks, and better anticipate how future quantum computers will be able to solve new and more challenging problems.

In this chapter, we set out the basic principles of quantum mechanics and show how they affect the possibilities that quantum computing offers and the challenges that the industry faces in bringing those possibilities to life.

TIP

Quantum mechanics is familiar to many people from college courses or from studying the many technologies based on quantum mechanics, some of which are described in Chapters 4 and 5. But many other people are unfamiliar with quantum mechanics, which has still not fully made its way into high-school and even some college science textbooks. So, if you are new to these ideas, you might consider reading this chapter twice. One of the authors (Smith) has been writing *Dummies* books for more than 30 years and doesn't remember ever giving this advice in one of his other books. But this may be one time it's worth considering.

HORATIO'S NEW PHILOSOPHY

In the play *Hamlet*, the title character asserts: "There are more things in heaven and Earth, Horatio, than are dreamt of in your philosophy." Perhaps Shakespeare, whose play was first published in 1603, had somehow received a glimpse into the future of physics. Determinism is an underlying principle of many branches of philosophy. But quantum mechanics clearly states that even the most basic operations of the universe — *especially* the most basic operations of the universe — are subject to the laws of probability. In addition, not only do we not know what's going to happen until we observe it, but the very act of observation changes the reality we're observing.

But wait, it gets worse, if you're a determinist. (As most of us, by experience and education, are.) Quantum mechanics also states that particles are entangled, and that information about one can affect the state of another faster than the speed of light. The entire universe is, in a sense, a bowl of intertwined spaghetti, connected by entanglement.

These philosophical assertions are right at the core of quantum computing, dependent as it is on superposition (that is, probability) , entanglement, and tunneling — the possibility that a particle will suddenly appear on the other side of a barrier. Keeping their implications in mind can help inspire us to get the most out of the new possibilities that quantum computing offers.

TIP

For a thorough investigation of quantum mechanics as its own topic, please see the excellent *Quantum Physics For Dummies*, Revised Edition, by Steven Holzner. (We use the term *quantum mechanics* instead of *quantum physics* in this book; the two terms are more or less interchangeable.)

Identifying the Keys to Quantum Mechanics

To boil an extremely complex topic down to a few words, the keys to quantum mechanics are quantization, uncertainty, and coherence. These principles apply to fundamental particles — a somewhat fuzzy concept that we explain in the next section.

Briefly:

- *Quantization* means the universe is "grainy" — space, time, matter, and energy are made up of discrete, indivisible units, much like tiny grains or particles that can't be subdivided. For instance, the electrons circling the nucleus of an atom can be in any of four orbitals around the nucleus, representing from one, two, three, or four units of energy each.
- *Uncertainty* means that we can't know both the location and momentum of a fundamental particle at the same time; the more closely we measure one, the more uncertain we are about the other. As a result, electron orbitals are usually represented as clouds around the nucleus, reflecting our knowledge as to where the electron might be and the uncertainty around its exact location at any specific time.
- *Coherence* means that a fundamental particle is free of interaction or measurement, and therefore able to support superposition and entanglement (described later in this chapter). Protecting each qubit from noise — such as vibration, heat, or changing electromagnetic fields — keeps the fundamental particle at its core coherent.

REMEMBER

Qubits are the processing units that power a quantum computer. At the core of each qubit is a tiny amount of coherent matter, though that core can be as small as a single photon, or an entire atom, or even a tiny superconducting piece of metal made up of many, many atoms. (Superconductivity allows the metal to stay coherent despite its many atoms.)

Figure 3-1 is a visualization of electron orbitals, drawn to represent the most likely locations of the electron. An electron in a given orbital — that is, at a given level of energy — might also, with diminishing probability, appear farther, even much farther, from its nucleus.

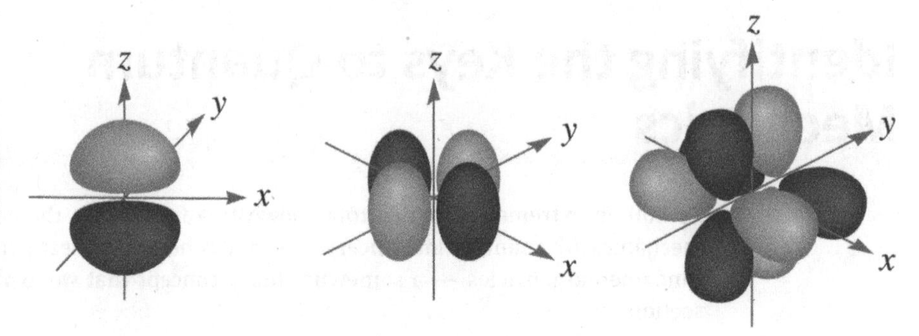

FIGURE 3-1: Electron orbitals are rendered as shapes that show the most likely position of an electron in each orbital around an atom's nucleus.

Finding the fundamentals of fundamental particles

Quantum mechanics and quantum computing are so different from our daily experience that we have to use words in new ways to describe what's happening. *Coherence* and *fundamental particles* are two of the most important terms in our quantum computing vocabulary.

A *fundamental particle*, also called an elementary particle, is matter that can't be subdivided further into other matter. A fundamental particle can be put into a state of coherence and will remain in that state until it's measured or until it interacts directly with energy or with other matter. (*Until it's measured* and *until it interacts* mean roughly the same thing here — measurement is an interaction done with conscious purpose, and interactions force a particle to, in effect, be measured by another particle.)

An electron is a fundamental particle; a human body, made up of many fundamental particles that are constantly interacting with one another, is not. Protons and photons are made up of smaller particles but act as single particles for most quantum mechanical purposes. Even entire atoms can display quantum mechanical behavior. To oversimplify, the smallest units of matter — photons, protons, electrons, and atoms — can be kept coherent; most larger units cannot.

What's both confusing and exciting is that some rather large assemblages of matter can also be kept coherent, usually under superconducting conditions. To achieve superconductivity, you must first select the right kind of matter — usually a specific metal or gas — and maintain it at an extremely low temperature, within a degree or so of absolute zero.

In a supercooled state, the atomic nuclei vibrate far less, due to the lower temperature. The electrons are largely freed from their nuclei, allowing the superconductor to carry electric charges with virtually no resistance. (The name *superconductor* breaks down to *super*, meaning *very*, and *conductive*, meaning *able to convey an electric charge.*)

The largest superconductors in the world are the superconducting electromagnets used to propel magnetic levitation (maglev) trains in Japan and China. Superconductivity enables a very strong magnetic field that has propelled trains on a test track at up to 600 km/h (nearly 400mph). Figure 3-2 shows a mag lev train in Japan.

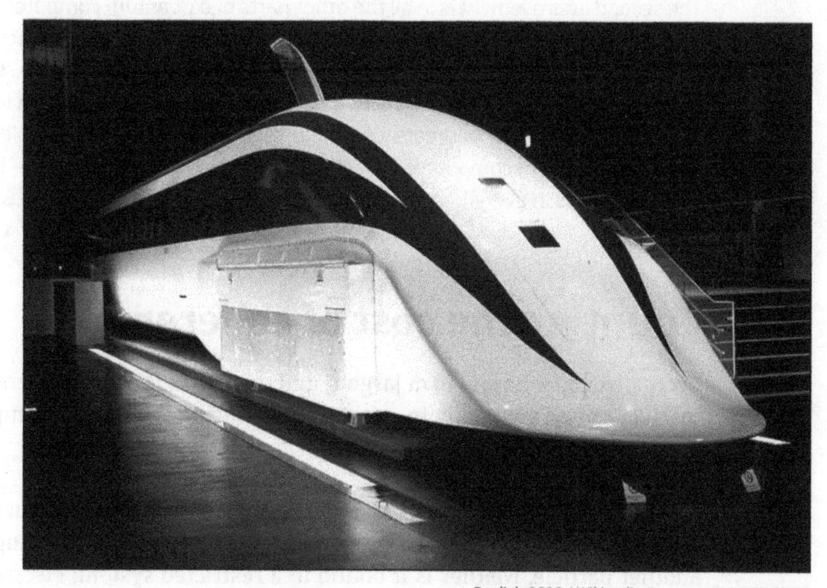

FIGURE 3-2: A superconducting maglev train running on the first operational line in Japan.

Many of today's quantum computers use qubits made from tiny loops of metal kept at superconducting temperatures. Other qubits are made from the smallest units of matter, including electrons, photons, ionized atoms, and neutral atoms. See Chapters 10 and 11 for a longer list.

Extending superconductivity to gases, a *Bose-Einstein condensate* (BEC) is a gas of a specific chemical composition kept at very low temperatures, enabling superconductivity. BECs are used as qubits in the lab, though not yet in any commercial quantum computers.

TECHNICAL STUFF

When a Bose-Einstein condensate explodes, it's called a bosenova. Seriously.

BUILDING QUANTUM COMPUTERS WITH QUANTUM COMPUTERS

New computers are designed on existing computers. One of the first purposes for quantum computers will be to design better, faster quantum computers.

This potential improvement has two sides. The first improvement is in materials science. Quantum computers are expected to become really good at simulating quantum properties in new materials, and a prime target for improvement in materials is a cheap, easy-to-manufacture superconductor that operates at higher temperatures.

The second improvement is in all the other parts of a quantum computer. Everything from controls for managing qubits to quantum algorithms may be subject to improvement as quantum computers themselves improve. For instance, a more stable superconducting qubit, perhaps using a far smaller chunk of metal, could be a winner. This recursive cycle of improvements yielding further improvements, ad infinitum, has barely begun for quantum computers. When it gains momentum, the results — in quantum computing, and in everything else that people do — are likely to amaze us all.

Counting the cost of coherence

Every discipline has its own jargon, and the use of the word *coherence* in quantum mechanics and especially in quantum computing is a prime example. Here's a try at defining it.

Coherence is the condition of an atom or a subatomic particle, such as an electron or a photon, that is not being measured and is not interacting strongly with another particle. Neither is it bound in a restricted system; electrons in the inner or closed orbitals of an atom, for instance, are bound in a restricted system. In some cases, such as certain supercooled metals, a chunk of material that is not being measured and is not interacting strongly with one or more particles can also demonstrate coherence. The coherence of some supercooled metals is caused by their having free electrons that are not bound within atoms and are free to move within the material, due to its being a superconductor.

While coherent, a particle — defined here, circularly, as any item of matter that can display coherence — displays quantum mechanical properties. It's subject to the *uncertainty principle;* that is, we can't measure both its location and its momentum (mass and velocity) precisely. And it can exhibit superposition and entanglement and tunnel through barriers.

Quantum computing depends completely on coherence. The core of a qubit is a particle kept in a state of coherence so that it can be kept in a state of superposition, maintain entangled connections to other coherent particles, and in some cases tunnel in ways that help its operators simulate quantum properties or solve computational problems.

But coherence has a cost. To keep a particle coherent, quantum computer operators must keep it free from measurement, interaction, or interference. Doing this reliably, even for brief periods, requires cutting-edge engineering.

Most qubits are kept coherent for only fractions of a second. In this window of time, operators program a set of qubits by gently and repeatedly manipulating them using laser beams, microwaves, or magnetic fields, and then read out the results.

When either accidental interference or purposeful measurement takes the particle at the core of a qubit out of coherence, it's said to *decohere*. If the decoherence is caused by a purposeful act of measurement, operators get a 0 or a 1 out of the qubit at that moment. Then the operators must make the qubit coherent again before it can be reused.

Identifying the Effect of Uncertainty

The Heisenberg uncertainty principle is crucial to quantum computing. Qubits are made up of fundamental particles kept in a state of coherence. As long as a particle stays coherent, crucial properties of the particle can hold many values simultaneously.

The property used in most quantum computing work to date is the spin of the particle. When a computation is complete, the spin of each qubit is measured, which causes the qubit to decohere and yields a 0 or 1 as a result.

TECHNICAL STUFF

As you might have suspected by now, a particle's spin is not the same as the spin of a planet or a baseball. It's a convenient term that describes the way the particle's magnetic field interacts with its surroundings. For quantum computing purposes, we describe the *spin* of a particle as being either up or down, and by convention we assign a value of 1 to up and 0 to down.

Because quantum computing is probabilistic, each run of a program might produce a different result for each qubit in the quantum computer. Operators have to run the program many times to determine the probability of obtaining a 0 or a 1 from a given qubit for each run of the program. This yields a statistical distribution of outcomes, which can be used to infer the probability of each possible outcome. The overall group of qubits will yield a different string of 0s and 1s on each run of the program.

By executing the program multiple times and measuring the qubits each time, operators obtain a large number of output strings. These strings can be used to construct a probability distribution which, by providing insight into the likelihood of each possible output, serves as the result of the program.

The role of quantum mechanical uncertainty is muddled by the fact that today's qubits are vulnerable to errors and lack error-correction capability. As a result, many errors appear in quantum computer runs. It takes many program runs and the use of sophisticated statistical techniques to factor out much — not all — of the errant results and obtain a useful final result.

Uncertainty also applies to *entanglement*, the connection between two fundamental particles that links properties such as momentum and spin. For a given particle, we don't know its spin — in quantum philosophical terms, it doesn't have a specific spin — until we measure it.

But if the particle we measure is entangled with a second particle, measuring the spin of the first particle instantly tells us the spin of the second particle, even if the two particles are very far apart. (This holds true even if one particle is on Earth and the other is in space, as shown in a recent Chinese experiment.)

The fact that qubits can be entangled, and that the entangled connection is itself in a state of superposition, exponentially increases the power of quantum computing. The fact that entanglement operates in a faster-than-light (FTL) manner greatly increases the potential speed of quantum computing.

Summarizing the History of Quantum Mechanics

We've shown how quantum mechanical principles effect quantum computing. But how did quantum mechanics develop? And what does this history tell us about the possibilities for quantum computing?

Tracing the development of classical mechanics

First we should paint a picture of how the world looked before the birth of quantum mechanics. The mathematical underpinnings of physics, called *mechanics*, have evolved over several centuries. *Classical mechanics* is the name for the system developed and used before quantum mechanics. The non-quantum computers in use today are based on classical principles, so we refer to today's mainstream computing as *classical computing.*

Copernicus, Galileo, and Newton were the leading figures in the development of classical mechanics. Before Copernicus, European scientific thought was developed in partnership with the Catholic Church, so scientific and religious ideas and ideals were intermixed. New scientific theories were subject to scrutiny and, in many cases, harsh punishments from religious authorities.

Before Copernicus, Europeans believed that the Earth was at the center of the universe, with planets and stars circling around it. (Some other cultures had figured out the truth long before.) This is referred to as the geocentric (literally, *Earth-centered*) model and also as the Ptolemaic system, after the Greek philosopher Ptolemy, who set out this view in about 150AD. The fact that planets could often easily be seen moving backwards across the night sky made this notion difficult to support.

Just a few months before his death in 1543, Copernicus theorized that the planets revolved around the Sun, publishing his findings in the book *On the Revolution of the Heavenly Spheres.*

Copernicus's publisher forestalled the wrath of the authorities by including a note in the book stating that the theories in the book need not actually be true — but they led to better calculations by astronomers, so they were useful even if they were false. Many proponents of quantum mechanics and quantum computing argue that their theories and implications are useful for practical applications, regardless of whether we fully understand or believe in them. This approach is sometimes referred to as "shut up and calculate," reflecting a focus on practical results and an attempt to stay out of philosophical debates.

Galileo published his own ideas in a book called *The Dialogue Concerning the Two Chief World Systems* (see Figure 3-3), which compared the geocentric and Sun-centered, or heliocentric, views of the planets. It was eventually banned by the Catholic Church.

Galileo used a telescope — invented by others, but improved by himself — to search the heavens, discovering the moons of Jupiter. From the Jovian system and other observations, he concluded that Copernicus had been correct: The Sun is at the center of the solar system and the Earth is one of several planets circling it.

Galileo published his ideas in 1632. As a result, he was put on trial and found guilty of heresy. Under threat of execution, he was forced to publicly recant his beliefs, and spent the last ten years of his life under house arrest. He published one final book just before he went completely blind then died in 1642, aged 77.

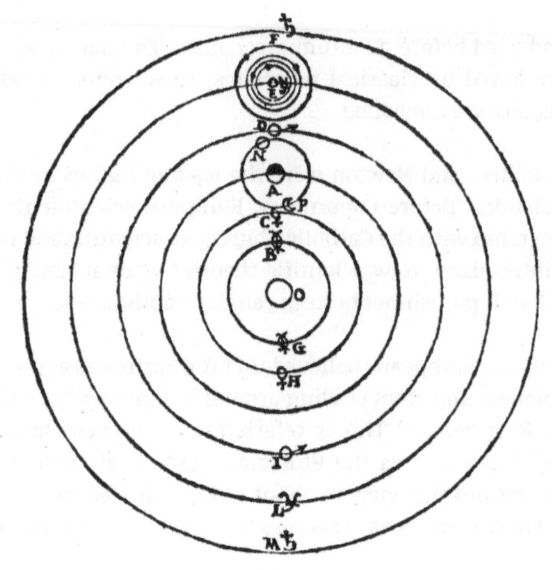

FIGURE 3-3: Galileo's *Dialogue* presented his heliocentric view of the planets.

Image Asset Management / AGE Fotostock

But the heliocentric cat was out of the bag, and philosophers, mathematicians, and physicists used the new ideas to develop new theories. Chief among them was Sir Isaac Newton, who published his *Principia Mathematica* in 1687. He used mathematics to describe the movements of heavenly bodies and the workings of space and time. He developed and used calculus, which begins with the summation of infinitesimal differences between quantities and then goes much further.

Despite his genius and immense accomplishments, Newton was still involved in a number of debates. He and others asserted that light is made of tiny, indivisible particles. Christiaan Huygens and others asserted that light is a wave moving through the ether, a mysterious substance that was said to pervade all of space. For centuries after Newton's death, the wave theory predominated. But it would turn out that Newton and Huygens were each partly right.

Science developed rapidly as a result of the work of these and other giants. The technology in use at the time of the American Revolution in the 1770s would have been easily recognizable to the Greeks and Romans — travel by horse and horse-drawn carriage (on roads far worse than Roman roads), manufacturing powered by waterwheels, and communication by handwritten letters, often in Latin.

But manufacturing began in England in the 1760s, at around the time of the American Revolution. By the end of the 1800s, the world had the steam engine, the cotton gin (for milling cotton without painful handwork), trains, metal ships, and the telegraph. The US Civil War of the 1860s was the first modern war, with horrors such as troops with rifles charging positions defended by Gatling guns, an early version of the machine gun.

TAKING A STEP TOWARD MODERN COMPUTING

During the mid-1800s, the famous mathematician Charles Babbage worked on his Analytical Engine, a proposal for an advanced mechanical computing device. However, the device could not be built because manufacturing technology at that time could not reliably produce gears to the required tolerances.

Ada Lovelace, a British mathematician and writer, developed an algorithm for calculating a sequence of Bernoulli numbers on the Analytical Engine, and this algorithm is widely regarded as the first computer program.

She described a vision for the future of computing that included machines that could create art and music, all of which has made her an iconic figure in computing. The Ada computing language, originally developed for the US Department of Defense in the late 1970s, is still in use today.

Also during this time, great progress was made in chemistry and biology, including the creation of the periodic table of the elements, the publication of Darwin's theory of evolution, and the emergence of the germ theory of disease. (Penicillin, though, was not discovered until 1928.)

The defining characteristic of science up to this point is that processes were seen to be continuous. A thrown ball moves smoothly through space, slowed gradually by friction from the air; a heated object gradually cools to meet the ambient temperature around it.

This continuous world was believed to be predictable. If you knew the initial state of any system, including the entire universe, you could in theory predict its later state, with as much precision as you had the resources to use in your calculations.

In a world where town centers were dominated by clock towers, so that everyone could easily find the correct time, this predictability was called the *clockwork universe*. The laws of physics meant that scientists could run the universe forward or backward, accurately, from any given point of time. This vision of reality was soon to be upended by quantum mechanics.

Beginning the quantum revolution

The theoretical underpinnings of the quantum revolution were developed from about 1900 to 1930 and laid the foundation for so much of what makes up our modern world, which continues to change as quantum mechanical ideas and ideals are applied ever more broadly.

The application of quantum mechanics to technical progress has resulted in transistors — the basis of all the classical-computing power we leverage today — lasers, display screens, medical scanners, and so much more. From the smartphone in your pocket to the TV in your living room to the fiber optic cables that carry internet traffic, the modern world is unimaginable without the quantum revolution.

Quantum computing, as important as it is, is just the latest major application area for the results of this wholesale change in science, mathematics, and even philosophy.

The evolution of quantum mechanics is often divided into two parts. The first quantum revolution is the development of quantum mechanics and its application to technology in a semiclassical fashion, without the capability to directly manipulate individual particles while in a state of coherence. Some of the main kinds of devices developed as part of the first quantum revolution are described in Chapter 4. The second quantum revolution is the development of quantum information science and the creation of quantum information theory, quantum computing, and quantum cryptography, all enabled by the devices created in the first quantum revolution. The second quantum revolution is described in Chapter 5 and after.

The quantum revolution began with an anomaly in the radiation of energy from heated objects, such as a stone or a block of cast iron. Theory said that such radiation would be high frequency, such as thermal energy and infrared light. But experiments showed that energy was emitted at a range of frequencies, including visible and even ultraviolet light. This became known as the *black body problem.*

In describing the progress of quantum mechanics during the core early decades of its development, we have included a number of equations. These equations are iconic and important, but you don't need to understand them to understand the progress that they made possible. Feel free to skip over the equations if you find them confusing or distracting.

Max Planck, investigating the issue in Germany in 1900, came up with a formula to describe the range of radioactive waves that a heated body emitted:

$$E = hf$$

E is the energy, f is the frequency of the radiation, and h is a very small number now known as Planck's constant, which relates the energy of a photon to its frequency. (The energy of the photon can be converted to its mass using Einstein's famous formula for mass-energy equivalence, given shortly.)

Planck's constant is

$$6.62607015 \times 10 - 34 \text{ m}^2\text{kg} / \text{s}$$

This extremely small number is the fundamental unit of quantization. It describes the inherent "graininess" of the universe.

Planck posited that energy, rather than being emitted across a continuous range of frequencies, is instead emitted at discrete, fixed frequencies separated from each other by Planck's constant. These specific frequencies represent the only values that can be emitted, with no frequencies existing between them.

Planck did not like quantization and spent years trying to backfit his (brilliant) findings onto the previously existing, continuous, deterministic structure of classical physics. His brave labor only served to prove that the new approach he pioneered is true and laid the groundwork to show that it applies more broadly than anyone at the time imagined.

In 1905, just a few years later, German patent office clerk Albert Einstein wrote several epochal scientific papers. One described special relativity. Einstein asserted that the speed of light, 186,000 miles per second, is the fastest speed in the universe. Nothing, including information, can move faster than the speed of light. This principle is called *locality*. Einstein also showed that two observers moving very quickly toward or away from one another can each still use light as an absolute yardstick for speed.

In the special relativity paper, Einstein set out the single formula that may have done more than any other to change our understanding of the universe:

$$E = mc^2$$

In this equation, called Einstein's mass-energy relation, E represents the amount of energy contained in a given amount of mass. m represents the amount of mass. c is the speed of light — a large number, which is then squared, producing a very, very large number. At larger scales, or when matter is bound in a restricted system, such as a proton in the nucleus of an atom, the energetic nature of matter is severely constrained, and the matter acts as a solid object.

Among other applications, this formula pointed the way to atomic energy and nuclear weapons. The amount of energy that can be liberated by converting mass completely to energy is enormous. For example, converting the matter in a paper clip entirely to energy would liberate the same amount of energy as one of the early hydrogen bombs.

In another paper, Albert Einstein used Planck's ideas and his own new thinking to come up with a radical explanation of the nature of light. He asserted that light is simultaneously made up of waves and solid particles. Shining of light at a metal plate can, for instance, cause the metal plate to become electrified, which is called the *photoelectric effect*.

MORE REASONS WHY EINSTEIN MATTERS

Einstein's mass-energy relation has a philosophically stunning meaning: Mass and energy are two facets of the same thing. Another way of describing this would be to say that mass is congealed or concentrated energy.

This way of thinking is helpful in understanding how qubits work. Matter in a state of coherence is acting more as bundles of energy than as tiny, solid billiard balls or similar everyday matter. It's easier to think of bundles of energy, which act as waves, as exhibiting superposition (not having a single defined position) or being entangled (energetically connected to each other).

By definition, these particles, for which he invented the name *photons,* always move at the speed of light. The energy of a photon is given by a simple formula:

$$E = hc / L$$

In this formula, E is the energy of a photon, c is the speed of light, L is the wavelength of light (which determines its color, for visible light), and h is . . . wait for it . . . Planck's constant.

Einstein's work demonstrates that light can be both a particle and a wave, showing different characteristics in different circumstances. Einstein won the Nobel Prize for this explanation of the photoelectric effect, with the Nobel committee making no mention of special relativity.

In 1913, Niels Bohr of Denmark applied Planck's ideas to the structure of the atom, resulting in a revolution in physics and chemistry. He asserted that atoms are made up of positively charged protons and uncharged neutrons, making up a nucleus, with negatively charged electrons in orbit around it. This idea was new and revolutionary.

The electron orbits are not at random distances, though, but quantized. A lowest-energy orbital can hold two electrons; a second, more distant orbital, one level up, can hold four electrons; and so on. The difference between orbits was given by Bohr's formula:

$$L = nh / 2\pi$$

That is, L, the angular momentum — energy of rotation — of the electron equals an integer, n, such as 1, 2, or 3, times h. (Yes, says the smart-aleck kid in the back of the room, Planck's constant again.) This is divided by 2π to account for the circularity of the orbit.

Among other things, Bohr's model explains spectral lines. When an element is heated, it emits colored light, with a different set of colors for each element. The distance between spectral lines depends on Planck's constant. The human visual system has evolved to assign colors to distinguish between the energy levels of the photons which reach our eyes.

Bohr's model gives us an explanation for the spectral lines emitted by different elements. Each time an electron jumps from one orbital to another, it emits a photon, whose energy is determined by the mass of the nucleus and the distance between orbitals — and we see the different photonic energies as distinct colors.

Continuing the quantum revolution

Planck, Einstein, and Bohr, along with many others, had established the first foundations of quantum mechanics by 1913. But then things got even stranger, in ways that are still changing the world today.

Einstein, like Bohr, was intellectually and emotionally attached to the structures and certainties of classical physics. He would spend much of his life arguing against some of the more challenging implications of his own and others' ideas. Einstein had shown that light behaves like a wave sometimes — for instance, if it passes through a narrow slit, it diffracts like a wave. But when a photon collides with an electron, the interaction can be fully described as a collision between two particles, without reference to waves.

In 1924, the Frenchman Luis De Broglie took Einstein's idea of the wave-particle duality of light and applied it to electrons. Electrons are not only particles but also standing waves. The wavelength associated with any object — yes, including your dome-like skull — is given by yet another simple formula:

$$\lambda = h / mv$$

λ is the wavelength, m is the mass, and v is the velocity. No points, by now, if you recognize h as Planck's thrice-bedeviled constant.

At speeds well below the speed of light, the same equation can be written as

$$\lambda = h / p$$

p is the object's momentum, which is simply a name for the product of its mass and velocity.

The de Broglie equation also tells us why we humans think objects are solid and are so surprised when tiny objects such as photons act like waves. The de Broglie wavelength of a baseball moving 80mph (130 kph) — a major league curveball — is about a trillionth of a trillionth of a trillionth of the size of the baseball. The wavelength is infinitesimally small compared to the size of the object it's attached to.

But at the subatomic scale, the de Broglie wavelength of a particle is significant and is on a similar scale to the particle itself. In this realm, both the wave and the particle nature of matter are important and cannot be ignored.

The trouble is that we have a great deal of difficulty visualizing these complementary aspects of reality in our mind's eye or describing them in words. They challenge our conceptual abilities and our language.

At very small scales — and even for some special macroscopic objects, such as the superconducting magnets that boost the speed of Japanese maglev trains — what we are looking for helps to determine what we find.

Although progress in quantum mechanics continued, describing it coherently was hard. (Tell us about it.) But in 1925, Werner Heisenberg, Max Born, and Pascual Jordan came up with *matrix mechanics*, a way of describing transitions between quantum states. This represents a big advantage in describing quantum mechanics, and matrix manipulation is at the core of quantum computing, as you will see in later chapters.

In 1927, Werner Heisenberg summed up the uncertainty implied by wave-particle duality in his famous uncertainty principle. The smaller an object, the larger the uncertainty in its momentum (again, its mass times its velocity).

The formula for Heisenberg's Uncertainty Principle is

$$\Delta x \, \Delta p \geq h / 4\pi$$

Δx is the standard deviation — a measurement of uncertainty — for an object's position, and Δp is the standard deviation for momentum. h is of course Planck's seemingly inescapable constant — itself a precise expression of the fundamental graininess of existence.

When you try to decrease the uncertainty for position, you increase the uncertainty for momentum, and vice versa. And this *conservation of uncertainty* is a fundamental characteristic of reality, not something we can mess around with or evade.

Heisenberg's Uncertainty Principle dethrones the clockwork universe. In a world where the uncertainty principle holds, nothing, at small scales, is certain. The past is out of reach (we can't calculate our way back to a precise description of it), the present is constantly surprising us, and there's no way to predict the future with any precision.

Thus, Heisenberg's Uncertainty Principle was and still is a source of great consternation to those who believe that the universe is obligated to be reasonable, predictable, and certain. Instead, the universe is fundamentally unknowable — past, present, and future.

To us, the authors, the wonderful thing about quantum computing is that it depends entirely on this otherwise frustrating and confounding squishiness. Each qubit is simply a little piece of the universe's uncertainty that we entangle, program, and interrogate to find out what may be possible.

SCHRÖDINGER AND THAT DARNED CAT

No actual cats were harmed in the writing of this sidebar.

In an attempt to highlight the ridiculousness of some of the quantum mechanical claims underpinned by his own work, Erwin Schrödinger described a thought experiment to bring the uncertainty that we find at the quantum level to the macroscopic world.

In his infamous thought experiment, a cat is placed in a sealed box with a small amount of radioactive material, a Geiger counter, a hammer, and a vial of poison. If the radioactive material decays, the Geiger counter will detect it and trigger the hammer to break the vial of poison, killing the cat.

However, according to quantum mechanics, the decay of the radioactive material exists in a state of superposition until it is observed; it is neither decayed nor not decayed. This means that until the box is opened and the cat is observed, the cat is neither alive nor dead but exists in a superposition of both states.

Now at this macroscopic scale where cats, hammers, and flasks of poison exist there are so many interactions above the quantum level that superposition can't last for more than a few femtoseconds. A Schrödinger cat will actually live or die at some specific time, and the experimenter can measure its liveliness or rigor at any time.

But ironically, the seeming contradiction at very small scales that Schrödinger wanted to highlight is, indeed, true, as verified by a century's experimentation in quantum mechanics.

And, it is still not true that any cat was harmed in the creation and execution (ha ha) of Schrödinger's thought experiment. It is also not true that the technical term for joy in the discomfiture of cats is Schrödenfreude. (In case that joke falls flat for you: The German term for *joy* in the discomfiture of others is *schadenfreude*.)

Laying the groundwork for quantum computing

Even with all the advances up to this point, the atom still had more secrets to give up. In 1926, Schrödinger came up with an equation describing the wave function of a quantum mechanical system, and he applied it first to hydrogen atoms.

The equation uses calculus and notation that many readers will not be familiar with and that are hard to describe in plain language. To summarize, Schrödinger's equation says that the position of a subatomic particle, such as an electron, is a wave function. The wave function varies with time. Initially, this was assumed to mean that the position of the electron, as a solid particle, varies with time.

German physicist Max Born provided a new interpretation of Schrödinger's equation that focused on the probabilistic nature of quantum mechanics. Instead of describing the precise position of a particle like an electron, the equation gives the probability that the particle will be found in a specific location. This probability is described by a wave function that changes over time, meaning that the probability of finding the particle in different locations also changes.

In this interpretation, which seems after a century of experimentation to be correct, there is no fixed object in Schrödinger's equation — no electron, for example. There's simply a probabilistic description of where we are more or less likely to find an electron if we look for it. Taking the measurement, in a sense, summons the electron into being at one place or another.

This interpretation drove many more traditionally minded physicists, not least Schrödinger himself, to distraction. Schrödinger had a nervous breakdown out of frustration over the implications of his own work, even as more uncertainty-minded colleagues such as Born celebrated the same implications.

The wave function is a superposition of possible states — in this case, locations — of the electron. This is the same superposition that we use to assign to different potential results for a quantum computing program.

Einstein joined with Schrödinger in pushing back against the idea of uncertainty as a fundamental property of the universe, objecting deeply to the idea that only observation causes things to manifest into reality out of clouds of possibilities. Born and others embraced the uncertainty that had been called out by Heisenberg in his Uncertainty Principle, and it's this more open-minded group who have made all the running in the many experiments that test the assertions of quantum mechanics.

These debates came to a head at the Fifth Solvay Conference in 1927, which had quantum mechanics as its focus. Figure 3-4 shows a photo of the attendees.

FIGURE 3-4: The fifth Solvay conference included most of the key contributors to quantum mechanics.

Benjamin Cobbery / Wikipedia Commons / Public Domain

The culmination of the initial burst of creativity that established the foundations of quantum mechanics may have been the publication, in 1930, of the first textbook on the topic, *The Principles of Quantum Mechanics*, by the famed scientist Paul Dirac. This textbook is still in use today.

Spotting "spooky action at a distance"

The most important work establishing the fundamental principles of quantum mechanics was completed by about 1930. However, new work based on these principles and arguments about what they mean continue to this day.

The pushback against quantum mechanics as a complete theory comes to a head around the idea of entanglement. Quantum particles do not have to collide, like two billiard balls, to interact. They simply have to come close enough to one another in spacetime for their wave functions — the ones described by Schrödinger — to overlap to a significant degree.

ASSEMBLING AT THE FIFTH SOLVAY CONFERENCE

The Fifth Solvay Conference, held in Belgium in 1927, assembled most of the key participants in the development of quantum mechanics in one place. It also provided an historic opportunity for two views of quantum mechanics to be hashed out in public view.

Planck, Bohr, and Einstein, considered to be the three founders of quantum mechanics, were all present and active at the conference. Einstein agreed that quantum theory is useful for describing nature at the atomic and subatomic level but expressed doubt that it was foundational to the whole of physics. Despite the assertions of quantum mechanics, Einstein maintained his conviction that the nature of reality is independent of the observer; he is not known to have ever changed his view on this question.

Niels Bohr served as a champion of quantum theory, pointing out that quantum interactions can't be observed directly but asserting that even indirectly observing them changes their results. And he pointed out that the probabilistic predictions made by quantum mechanics describe reality accurately, which continues to be true today.

Once the quantum particles interact, they become *entangled*. Key properties of one particle become tied to the same properties of the other.

The tricky part is that the particles remain entangled even as they travel great distances from each other, such that measuring a property of one entangled particle — which, according to quantum mechanics, causes that property's value to manifest — immediately tells us the corresponding state of the other. And this truth asserts itself instantaneously, with no limit imposed by the speed of light.

In physics, the idea that anything can move faster than the speed of light, including information, is called *nonlocality*. Nonlocality was, in theory, banned by Einstein's work on relativity dating all the way back to 1905. He called any violation of locality "spooky action at a distance."

In 1964, Irish physicist John Bell proposed ways to test whether nonlocality truly holds. Tests on the Earth and in space have confirmed that entanglement operates beyond the bounds of locality. Bell never won a Nobel Prize in physics, but he was awarded the Wolf Prize in Physics in 1988 for his contributions to the field. And three physicists who verified his work, in experiments conducted in the 1970s, 1980s, and 1990s, shared the Nobel Prize in Physics in 2022.

Today the implications of quantum mechanics extend beyond the world of physics and have sparked an intense philosophical debate. The field of quantum philosophy has emerged. This fascinating new area of philosophy explores the implications of quantum mechanics on our understanding of reality, consciousness, and the nature of existence.

For example, superposition and entanglement, two of the key concepts in quantum mechanics that we discuss in this chapter, have sparked heated philosophical discussion. Superposition suggests that quantum particles can exist in multiple states simultaneously until they are measured, challenging our traditional understanding of the nature of existence and the role of the observer.

Entanglement, on the other hand, enables quantum particles to remain connected and behave in a correlated way even when separated by large distances, raising questions about the fundamental nature of space and time, and raising questions as to whether everything in the universe is instantaneously connected at the subatomic level.

The philosophical implications of these quantum phenomena are still being explored and debated, and they have opened up new avenues for inquiry into the nature of reality as we face the limits of our current knowledge of the universe.

Despite the controversies surrounding these ideas, superposition and entanglement are at the heart of quantum computing. By exploiting these quantum properties, researchers are developing new technologies with the potential to revolutionize fields such cryptography, drug discovery, and materials science. It's remarkable that the same quantum phenomena that challenge our understanding of reality and have sparked philosophical debate are also the foundation stones on which quantum computing is built.

» Observing the evolution of electron microscopes

» Calling time on mechanical clocks

» Understanding the effect of the first wave

Chapter 4

Introducing Quantum Technology 1.0

Quantum mechanics, which emerged in a wild burst of creativity, confusion, and conflict between 1900 and 1930, led directly to two massive waves of technology. The first extended from roughly 1930 to 1980; the second began in about 1980 and continues to the present day. We can call these two waves of technology Quantum Technology 1.0 and Quantum Technology 2.0.

In Quantum Technology 1.0, scientists and engineers used the insights from quantum mechanics at a coarse level, working with large numbers of photons, electrons, and other quantum particles. To the extent that individual particles were considered, it was to get them to behave as part of the group and to not leak, tunnel, and so on.

Researchers initially worked at this larger scale because they did not have the tools to work with subatomic particles individually. The tools needed for Quantum Technology 2.0 were developed as part of Quantum Technology 1.0, and now these tools are becoming increasingly powerful.

In this chapter, we describe many of those innovations and we tie them to quantum computing, which is at the core of the wave of quantum technology we're living in now.

Finding Lasers at the Cutting Edge

Consider, for instance, a laser. A laser excites a bunch of atoms, synchronizes their oscillations, and then sends the resulting synchronized photons in bursts at some target, such as a growth on your cheek. Doctors (or, for larger growths, surgeons) don't aim individual photons; they send a beam of them. Nor do they attempt to aim at individual molecules or even cells. If a few photons go astray, that's a leak, not a potential source of computational power (as it is in quantum computing).

Doctors use a medical-grade laser (see Figure 4-1) to send the laser beam, a stream of excited photons, at the area containing the growth. The laser beam is likely to also vaporize or kill some benign or normal cells around the growth, but doctors manage that problem; they can't yet eliminate it. (Doctors do have better tools these days that they use for some procedures, but the laser is still "the cutting edge" for routine work.)

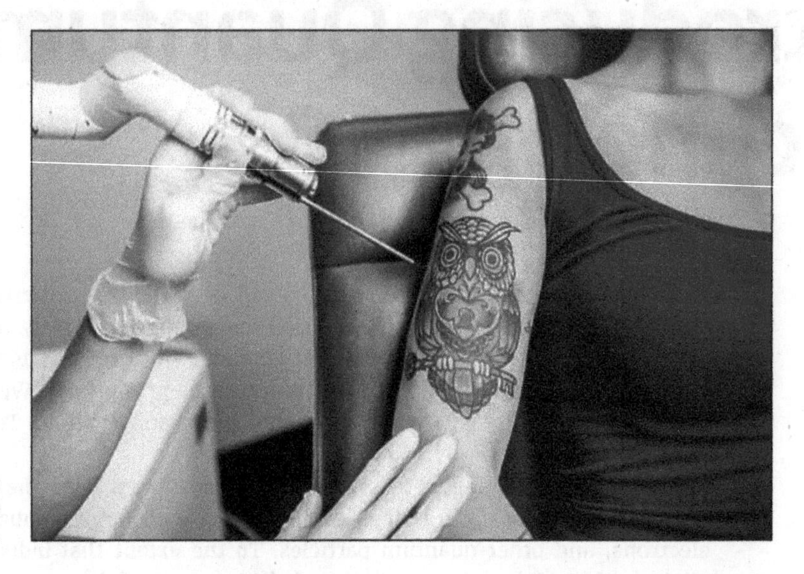

FIGURE 4-1: A medical-grade laser.

This general approach is characteristic of the first quantum revolution technologies that we describe in this chapter. As the laser example described here shows, during the Quantum Technology 1.0 wave of change, technologists worked, in a certain sense, wearing oven mitts. The tools did not yet exist to manipulate individual atoms and subatomic particles, such as electrons and photons. These particles could be managed and influenced in a group but not controlled individually.

THE MYSTERIOUS OBSERVER EFFECT

As scientists gained greater control of quantum technology, they lost control of the narrative regarding the nature of reality.

In 1801, Thomas Young created the double-slit experiment, in which a light was shone through slits in a metal plate. The light cast a shadow on a second plate that had been placed behind the first one. And this light formed a pattern with peaks and valleys — a wavelike interference pattern. The double-slit experiment seemed to show that light was made up of waves.

But then scientists gained the ability to send light through the slits one photon at a time. Although the photons were regarded as solid, they still caused an interference pattern to build up on the second plate — unless a detector was placed at the slits. If the photons were observed in this way, they went straight through the slits, forming two solid bars of light on the second plate.

The effect of the observer on the behavior of light remains a great mystery, with no solid explanation — but lots of hand waving.

The technologies developed in this wave also acted as the servants of classical mechanics, not subversive change agents that challenge people's grasp on reality. A laser is largely a superior knife, signaling device, or yardstick; an atomic clock is a much more accurate clock. These technologies don't leverage properties specific to quantum mechanics, such as superposition, entanglement, and tunneling.

TECHNICAL STUFF

Yes, we know that tunneling electron microscopes use, well, tunneling, but these microscopes, as cool as they are (they're supercooled, so pun intended) use electrons in relatively large numbers and depend on their overall predictability, not their individual capriciousness.

Studying Quantum Mechanics After 1930

When we left off in Chapter 3 on the story of quantum mechanics, the foundations of the field had been laid, and Paul Dirac had written the first textbook on quantum mechanics, published in 1930. And then what?

Quantum mechanics continued to be developed and refined. Also, the first wave of technological change driven by quantum mechanics began. And scientists and engineers began to use the principles of quantum mechanics to invent and improve things. Lots of things. But the work was challenging, and progress was measured in decades.

SHE BLINDED ME WITH (REVOLUTIONARY) SCIENCE

TECHNICAL STUFF

In 1982, before half of y'all were born, an artist named Thomas Dolby released a song, "She Blinded Me with Science." (Yes, it's on YouTube, if you're interested, and yes, it's in color.) As the song helps to demonstrate, science is indispensable to our world today. So much so that there's a branch of epistemology, the theory of knowledge, dealing with the nature of changes to scientific theory. The key book in the emergence of this theory is *The Structure of Scientific Revolutions* by Thomas Kuhn, first published in 1962.

Kuhn's core assertion, now widely accepted, is that scientific theory does not advance smoothly. Instead, a set of related theories and beliefs, often including religious and philosophical beliefs, become widely accepted — a little-questioned norm. Scientific work in these periods is called *normal science* and consists of incremental improvements to an accepted body of theory and practice.

When change comes, it arrives not as an adjustment, nor as a shift, but as a philosophical earthquake, upending people's view of the world. These are periods of *revolutionary science*. And, like other kinds of change, revolutionary scientific change is criticized and resisted, often fiercely, even violently, as we saw in the change from belief in an Earth-centered universe to a heliocentric worldview described in Chapter 3.

Actual violence often accompanies, or results from, revolutionary scientific change. Ludwig Boltzmann, the namesake for the Boltzmann constant in Planck's equation (Chapter 3), committed suicide in 1906 in the belief that his work supporting atomic theory was held in derision by his peers. (Atomic theory is now universally accepted, and Boltzmann's name and work live on.) And the discovery of quantum mechanics led directly, within a few decades, to the creation of atomic and nuclear weapons — violent devices indeed. This story is told in the hit 2023 movie *Oppenheimer*, which mentions quantum mechanics at several points.

The evolution from classical mechanics to quantum mechanics is wrenching and still incomplete. It may be that only children born in a future era, one in which quantum computers are in widespread use, will be able to fully accept these theories, which are now a century old but still seem new. And our children, with these new assumptions firmly in place, will dream dreams that those of us alive today can't imagine.

Here are the technologies we briefly describe in this chapter, along with dates that represent important early milestones for each of them:

- Solar cells (1890s)
- Electron microscopes (1931)

» Transistors (1947)
» Atomic clocks (1955)
» Masers and lasers (1953 and 1960)
» MRI scanners (1977)

Speeding the Race for Solar Cells (1890s)

Solar cells were invented in the 1800s, with the first patents granted in the 1890s. As with many similar advances, the inventors did their work without a solid understanding of the theory that lay behind it. It was Einstein's paper describing the photoelectric effect as a product of quantum mechanics in 1905 that began to provide the bigger picture.

It took many decades of advances in science and technology for solar cells to move forward significantly. Solar cells were produced for space activities in the 1950s by Bell Labs, reaching 2 percent efficiency in converting incoming light energy to electrical output. (Plants manage about 1 percent.)

TECHNICAL STUFF

Scientists have recently performed experiments showing that photosynthesis uses a quantum-computation-like process to transform sunlight into chemical energy that drives the life processes of plants (and animals, such as ourselves, that eat them). When a plant absorbs energy from sunlight, the energy moves through the photosynthetic system as a wave, exploring different possible pathways. The plants corral the energy into efficient pathways that create complex chemicals rather than inefficient pathways that waste the energy as heat. Scientists are eagerly investigating just how this works, partly in hopes of using the results to further improve the efficiency of solar cells.

DIVING INTO QUANTUM MECHANICS

There's a lot more to say about quantum mechanics, including ongoing work, than we can fit into this book. Luckily, a lot of good books — and, we fear, some less wonderful ones — describe this history and the current state of the art.

We recommend *Quantum Physics For Dummies*, Revised Edition, by Steven Holzner (Wiley, 2013) as an excellent starting point. If you want to do a deep dive, check out the accompanying book of exercises, *Quantum Physics Workbook For Dummies*, also by Steven Holzner.

Solar cells reached 10 percent efficiency in 1959, 20 percent efficiency in 1985, and 40 percent efficiency in 2006; the current record is just under 50 percent. Figure 4-2 shows new-generation solar cells used by NASA in space.

FIGURE 4-2: Advanced solar power cells energize space exploration.

NASA/JPL-Caltech/Lockheed Martin

Today, solar energy is cheaper than any other form of energy in many locations, and many different kinds of solar cells are available, including thin film solar cells and semitransparent (about 70 percent transparent) solar cells that can be used as windows.

Solar cells may save the world by helping people reduce the carbon emissions that reduce global warming, but they aren't used directly in quantum computing.

Eyeing Electron Microscopes (1931, 1965, and 1981)

Electron microscopes use electron beams rather than light beams to create an image with very, very high resolution — up to 100,000 times sharper than visible light. Electron microscopes were invented in 1931, just after Dirac's quantum mechanics textbook was published (see Chapter 3).

But electron microscopes had their limitations, particularly when it came to imaging surfaces. This led to the development of scanning electron microscopes (SEMs), which use a focused beam of electrons to produce detailed surface images. While the concept dates to the 1930s, the first practical SEM was developed in 1965.

Tunneling electron microscopes, first demonstrated in 1981, use the quantum mechanical phenomenon of electron tunneling to achieve even more precise measurements. With the tunneling electron microscope, atoms could be seen for the first time, not just inferred.

The number of electrons that tunnel through a barrier depends on its thickness. To determine the barrier's thickness, a sharp tip is brought extremely close to the surface of the sample being studied. Then voltage is applied between the tip and the sample, which creates a tunneling current of electrons in between. When the intensity of the tunneling current is measured, the data captured is used to build an image of the target surface. Like many other quantum devices, the accurate reconstruction of an image depends on a very high degree of isolation from possible sources of interference.

Electron microscopes and tunneling electron microscopes are useful in the design, manufacture, and operations of quantum computers, including the materials research used to build better qubits.

Optimizing the Transistor (1947)

Research into *semiconductors* — materials with limited conductivity that can be manipulated precisely — is a direct result of work in quantum mechanics. The first transistor was developed in 1947, and that work resulted in a Nobel Prize in Physics in 1956.

Transistors were quickly put to use in then-new electronic mainframe computers, ultimately replacing vacuum tubes — which are much larger, more power-hungry, and more failure-prone than transistors. (Fun fact: In the 1960s, convenience stores had testing machines people could use to troubleshoot the tubes used in their vacuum-tube-powered televisions.)

As we describe in Chapter 3, transistors and their descendants, the integrated circuit and the microprocessor (see Figure 4-3), are at the core of classical computing. Nearly every electronic device in use today is based on transistors, including computers, smartphones, smart watches, and any device with a screen.

FIGURE 4-3: A wafer of microprocessors.

Amazing studio / Adobe Stock

So quantum mechanical work is responsible for the rapid growth and current ubiquity of classical computing. However, as mentioned, transistors are carefully designed to avoid, not to leverage, quantum effects such as electron tunneling. So they are quantum mechanical devices that work in a classical mechanics framework. They use quantum mechanical insights to avoid quantum mechanical effects, enabling them to work more effectively, work more cheaply, work in a smaller physical space, and use less energy.

FROM TRANSISTORS TO TENSORFLOW

In 1958–59, two smart guys (and rivals) named Jack Kilby and Robert Noyce made something really cool. They called it the integrated circuit (IC for short). It changed the world forever, helping us make lots of things we use today such as TVs, iPhones, and even supercomputers.

Kilby and Noyce each made their own ICs, using different approaches, but both figured out how to put lots of electronic parts on a really small chip. Jack showed off his IC on September 12, 1958, and Robert introduced his version on July 30, 1959.

Their inventions made it possible to make really complicated electronic things on just one little chip. Only because of their work can classical computers support advanced applications such as Tensorflow, the leading machine learning platform. And in 2022, their innovative approach reached the quantum computing realm, with the announcement of the first quantum integrated circuit. Thanks, Jack and Robert!

TECHNICAL STUFF

The core quantum mechanical insight that drives many Quantum Technology 1.0 developments is that electrons exist in orbitals with different energy levels, emitting a photon when an electron moves to a closer, lower-energy orbital, and absorbing a photon to move out to a higher-energy orbital. This insight quantizes radiation — for instance, heat and light — which had been assumed in classical mechanics thinking to be continuous.

Telling Time with Atomic Clocks (1955)

Atomic clocks use the microwave signal emitted by an electron when it changes energy levels in an atom. The first widely used atomic clocks, developed in 1955, were built by Louis Essen and J.V.L. Parry at the National Physical Laboratory in London. These clocks depended on the magnetic resonance of cesium atoms to measure time. The original device was as tall as a person, but a chip-size version was later built and widely used.

TECHNICAL STUFF

Mechanical and electronic clocks have inherent inaccuracies — and who has time for that? Today, the most accurate way to get the time for most uses is to sync to an atomic clock over the internet, as your smartphone does on an ongoing basis. But there's variation in one's distance from the nearest available highly accurate clock and the speed of the internet between your location and there. So the most discerning users of time services either measure and account for these inaccuracies as best they can or purchase an atomic or quantum clock for onsite (or, in the case of vehicles, onboard) use, minimizing inaccuracy.

Over the decades, cesium clocks became more and more accurate. The global positioning satellite (GPS) system for location finding depends on atomic clocks — a quantum technology that most of us use every day. GPS was developed for military use and initially authorized for commercial use, as a safety improvement for airlines, in 1983.

In 2010, a new type of atomic clock, the quantum clock, was put into service, using ions cooled by laser beams. The new quantum clocks are far more accurate than the previous cesium standard. If a quantum clock had been set up at the beginning of the universe, nearly 14 billion years ago, it would currently be off by less than a second.

TECHNICAL STUFF

Quantum Technology 2.0, the wave we're in now, includes quantum sensing, a new field. *Quantum sensing* harnesses the quantum mechanical properties of matter to perform measurements that are more sensitive than any classical resource. Quantum sensing includes quantum clocks, powered by what can be described as single-purpose qubits, and current work toward a quantum GPS system, which will be way more precise than the current version. Figure 4-4 shows a quantum atomic clock.

FIGURE 4-4: A strontium ion optical clock.

Andrew Brookes, National Physical Laboratory / Science Source

Heating Up Masers and Lasers (1953 and 1960)

Maser- and laser-type devices were described in principle by Einstein in 1916, but working devices came decades later, from work in microwave spectroscopy. The first device using these principles, invented in 1953, was a MASER, an acronym for microwave amplification by stimulated emission of radiation. Shorter wavelengths are easier to manipulate, so microwaves — much higher in frequency than visible light — were the first type of radiation to be tamed in this manner.

An optical maser using light instead of microwaves, with the acronym LASER, was proposed in a paper in 1958, and a red light laser was demonstrated in 1960. (The terms *maser* and *laser* are now so widely used that they are no longer capitalized.)

Lasers have captured people's imaginations, though the death rays that many of us are quick to think of first have (fortunately) not seen widespread adoption. The first lasers used high-frequency infrared and red light; those of you who are old enough will remember the red LED screens of early portable computers. (These portable computers were not laptops, unless you had a lap of iron.) A great deal of work has been required to create practical lasers using higher wavelengths; a laser using blue light, the highest-frequency visible light, was not demonstrated until 1996.

TECHNICAL STUFF

When a laser beam of the right frequency is directed toward a specific type of atom, the light particle (photon) from the laser beam is absorbed and re-emitted by the atom. This causes the atom to lose energy, cooling it (see Figure 4-5). This technique is used for supercooling the qubits used in quantum computing approaches based on trapped ions and neutral, or cold, atoms, as described in Chapter 7.

FIGURE 4-5: A laser supercooling atoms.

Colorado State University

The focused heat produced by lasers has led to their use in surgery (both for cutting tissue precisely and, conversely, for "welding" tissues back together); for range finding and other measurements of distances, large and small; for compact discs (CDs) holding music and other data; and for laser pointers that are wonderfully effective for beguiling cats.

A critical use of lasers in today's connected world is for fiber-optic communications, the basis of the internet and the World Wide Web. Laser light is guided forward by the reflective walls of the fibers bundled together in fiber-optic cables. As a result of the use of laser-powered fiber optics, along with the use of transistors for classical computing of all types, today's telecommunications and computational infrastructure is unimaginable without quantum mechanics. Entanglement, a Quantum 2.0 technology if there ever was one, is being used to experimentally — err, quantumate? quantumize? quantify? — add quantum communications capability to existing fiber-optic networks.

TECHNICAL STUFF

Crews from *Apollo 11*, the spaceship that put the first men on the Moon, and later American and Soviet spaceships placed reflectors on the surface of the Moon and in orbit around it. Scientists frequently aim laser beams at these reflectors and use the round-trip time — about 2.5 seconds to cover the approximately 240,000-mile distance — to calculate the exact distance from the Earth to the Moon at a specific point in time. These measurements have shown that the Moon is very slowly drifting away from the Earth; the same measurements have detected wobbles in the Moon's location, which indicate that it has a fluid core.

As this example shows, lasers are important tools in scientific research as well as in quantum computing. In addition to supercooling qubits, lasers are used to trap and manipulate quantum particles in some types of quantum computers. Experiments with satellites and laser beams have also been used to show that entanglement — Einstein's "spooky action at a distance" — works faster than the speed of light. (Quantum communications in space! Very *Star Trek*, no?)

REMEMBER

Light is, broadly speaking, an electromagnetic field too, but the magnetic component is much less energetic and less important for most purposes, than the electrical one.

Scanning for NMR and MRI Devices (1977)

Nuclear magnetic resonance (NMR) is what happens when atoms are placed in a magnetic field and then an additional, oscillating magnetic field is applied. The nuclei of the atoms resonate in a measurable way that distinguishes atoms and molecules with different characteristics from each other. For instance, cancer cells hold a lot of water, so they stand out from normal cells on an MRI scan.

A magnetic resonance imaging (MRI) scanner is based on NMR principles. It builds up a three-dimensional picture of a substance — in the medical field, of living tissue — from these resonances.

MRIs are powerful and non-invasive diagnostic devices and can be used to diagnose, for instance, soft tissue injuries and cancers. The first whole-body MRI scanner, shown in Figure 4-6, was created in the 1970s by Paul Lauterbur and Sir Peter Mansfield.

TECHNICAL STUFF

The sharp-eyed among you will soon notice that we mention NMR here and NRM-based qubits in the quantum computing timeline in the next chapter. NMR-based quantum computers have qubits that use the principles of nuclear magnetic resonance to manipulate and control the quantum states of atomic nuclei, and qubits of this kind were at the core of the first useful quantum computers. So NMR technology, the last major innovation from Quantum Technology 1.0, is part of a bridge to Quantum Technology 2.0.

FIGURE 4-6: Sir Peter Mansfield with an original MRI machine.

Sir Peter Mansfield/The University of Nottingham

Assessing the Effects of Quantum Technology 1.0

Today's world is completely dependent on the technologies developed in the first wave of technological change made possible by quantum mechanics. You may at this moment be using energy from solar cells, delivered through the electrical grid or from a home solar array, to power a computer or tablet that you're using to read a digital version of this book.

A GPS sensor in your phone depends on atomic — and recently, perhaps, quantum — clocks to help guide you as you drive or walk to a nearby restaurant. The batteries in conventional, hybrid, and electric cars have been improved by the use of quantum mechanical principles. An MRI scanner may find, and a laser knife annihilate, a tumor that had been threatening your life. (Been there, done that, says author Smith.)

The quantum mechanics — powered technologies described in this chapter have powered a transformation of our world. However, as mentioned, they operate within the framework of classical mechanics. That is, they're used to more efficiently carry out processes derived from, and capable of being described by, classical mechanics.

The core strangeness of quantum mechanics, represented most strongly by superposition, entanglement, and tunneling, is not used directly in most of the first wave of technologies derived from quantum mechanics. But they're the very essence of the technologies that make up Quantum Technology 2.0, which we begin to describe in the next chapter.

» Using theory to fashion the future

» Making quantum computing universal

» Demonstrating real quantum computing

Chapter 5
Unveiling Quantum Computing

Quantum computing is the result of the fusion of early and classical computing (Chapter 2) and quantum mechanics (Chapter 3), aided by the use of the technologies developed as part of Quantum Technology 1.0 (Chapter 4).

All of us alive today are living in the time of Quantum Technology 2.0. In this continuing revolution, quantum information science (QIS), a new discipline, extends quantum mechanics in new directions. The result is three emerging disciplines: quantum sensing, quantum communications, and quantum computing, with quantum computing as the largest area of work.

In this chapter, we share a timeline of key developments in quantum computing to bring us from its origins up to the turn of the century. We're describing the beginnings of an earthquake while the ground is still in open revolt against gravity, so we're probably missing some key turnings, but please accept this as an early draft of the history that is still unfolding.

Quantum computing is shaping up to be the most consequential application of the principles of quantum mechanics to date. That's due to the fact that quantum computing uses the most amazing, even incredible, implications of quantum mechanics — superposition and entanglement — as foundational techniques.

The key results of this are

- Quantum computing is already changing the world — and may, in the coming years, change it to an even greater degree than the hugely consequential technologies described in Chapter 4.
- Qubits, though very much a work in process, are among the most incredible devices ever conceived of or built — the computational equivalent of having a nuclear reactor on a device smaller than a fingernail.
- Quantum computing also brings to life — and puts to work — the deep scientific and even philosophical questions at the core of quantum mechanics in a way that no previous technology has done.

Keep these larger implications in mind as we take you through key highlights of the emergence of quantum computing.

IMAGINING THE RESPONSE

The biggest early controversy in quantum mechanics, which continues today, is about the description of quantum particles, such as the photon, as waveforms with probability, rather than specificity, at their core. (Entanglement takes some more or less friendly fire along the way.)

Niels Bohr came to champion the probabilistic, non-deterministic point of view; Albert Einstein went to his grave as an advocate of the deterministic outlook. A century of experimentation, still continuing, supports Bohr's view.

It's intriguing to imagine the response of the pioneers of quantum mechanics to quantum computing, which depends to its core on the wilder assumptions of the hundred-year-old, but still shocking, probabilistic view.

For an insightful description of this ongoing debate, far beyond what we can address in this book, see the *New York Times* bestseller and Editor's Choice, *Quantum: Einstein, Bohr, and the Great Debate about the Nature of Reality* by Manjit Kumar (Norton, 2011).

Nailing Together a Framework for Quantum Computing

New theoretical work has resulted in the development of quantum information science. As mentioned, quantum information science includes three pillars:

- **Quantum sensing:** Using quantum mechanics to overcome the limits of current sensors
- **Quantum communications:** Using entangled particles for communication
- **Quantum computing:** Using entanglement and superposition for simulation and computation

This book focuses on quantum computing, but we discuss quantum communications and quantum computing as the topics arise. Developments in all three areas are included in the timeline that begins in the next section.

We leave out far more than we include, but it would take at least an entire book to give a reasonably complete history. The one area we are purposefully excluding from this timeline, however, is information about the evolution and use of different types of qubits; we address those in Chapter 6.

TECHNICAL STUFF

We could have dedicated a lot more space in this book to quantum communications and quantum sensing, but it would have come at the expense of a needed focus on the tremendous complexity and depth of quantum computing as it is emerging today. So our apologies to all of those doing breakthrough work in these exciting fields for not giving you more ink — or more pixels, for our readers engrossed in an e-book version — and we hope your work gets covered elsewhere in the depth that it deserves.

Theorizing in the 1960s and 1970s

As mentioned in Chapter 3, Ada Lovelace (see Figure 5-1) made significant contributions to the development of classical algorithms in the mid-1800s, before there was hardware to run them on. Charles Babbage had described the hardware, but no one at the time could machine parts precise enough to make his device work. (Perhaps the engineers who built the Antikythera — see Chapter 2 — might have helped.)

FIGURE 5-1: Augusta Ada King, Countess of Lovelace.

Artepics / Alamy Stock Photo

Similarly, work on quantum information theory and quantum algorithms often precedes the arrival of quantum computing systems to run them on. Following is a selective list of some early accomplishments in the relevant theory that have proven to be important to later work in quantum computing.

Foretelling quantum computing and nanotech

In a 1959 talk titled "There's Plenty of Room at the Bottom," Richard Feynman of Caltech asserted the possibility of using quantum mechanical principles for computation. He also proposed the new field of nanotechnology (now nicknamed *nanotech*), pointing out that biological systems read and write information at the atomic scale, including in DNA and RNA molecules.

Figure 5-2 shows Feynman giving his 1959 lecture.

FIGURE 5-2: Richard Feynman opens students' minds in a 1959 lecture.

The Big T / Wikimedia Commons / Public Domain

Keeping cool with reversibility

In 1961, Rolf Landauer, a German physicist who emigrated to the US in 1938, discovered Landauer's principle: Any irreversible operation that manipulates information, such as setting or clearing a bit, releases heat. Quantum information and quantum computing use this principle. Quantum computing uses reversible operations exclusively, so Landauer's principle tells researchers where not to go as they push these fields forward.

Quantizing money — it's a hit

Sometimes people are way ahead of their time. In 1968, after several years of work, the late Stephen Wiesner, then a graduate student at Columbia University, proposed quantum money. Quantum money was to be — and may someday become part of — the application of quantum physics to cryptography as a basis for financial services.

His work was so advanced that it was not accepted for publication at the time. It was finally included, as part of further work by colleagues, in a paper published in the early 1980s, and it hasn't yet been fully implemented in working systems, even all these years later.

Wiesner's ideas depended on using quantum states and the no-cloning theorem as a kind of quantum watermark on currency. His work was a forerunner of much of the most important work in quantum computing today, including Shor's algorithm for factoring large numbers on a quantum computer (described near the end of this chapter) and quantum key distribution for cryptography.

Cloning quantum states is a no-go no-no

In 1970, James Park published an article in *Foundations of Physics* proving mathematically that it's impossible to clone — that is, make an accurate copy of — a quantum state. This is part of a group of theorems called *no-go theorems*, which demonstrate that something (often something people would very much like to do) is not physically possible.

The inability to clone a quantum state has a big effect on quantum computing because copying information is highly desirable in many applications (and is fundamental to classical computing). However, as Park demonstrated, it's not possible in quantum computing, which is one of the mysterious aspects of the entire enterprise, along with superposition and entanglement. Park's work was later substantially duplicated by people who didn't know, at the time, of his early proof.

Much later, in 2000, the quantum no-deleting theorem was proven. Together, this theorem and the no-cloning theorem prove that quantum information can neither be created nor destroyed.

Regretting what you can't have

In 1973, Alexander Holevo shared some good news and some bad news. The good news was that a single qubit can carry more than one bit of information. The bad news was that you can't access anything beyond the first bit, and that the same limitation also holds for multiple-qubit systems. This upper limit is called *Holevo's bound*, and it's an example of an area where quantum computing is no more capable than classical computing. (In classical computing, individual bits are well understood to be capable of containing only a single bit of information, so that's all you can hope to access.)

Discovering a star (in information theory)

In 1976, Roman Ingarden of Nicolaus Copernicus University in Torun, Poland published the first paper on quantum information theory. Ingarden examined the foundations of quantum mechanics and the implications for information theory, laying the groundwork for quantum information theory.

Laying the Groundwork in the 1980s

With Benioff's paper on quantum Turing machines (described next), the prospect of real quantum computers began to get more serious. (Just as Turing's description of classical Turing machines in the 1930s led to working electronic computers in the 1940s.) Important work during this period included specifying the kind of logic gates that a universal quantum computer would need to do its work.

Extending Turing machines to the quantum realm

In 1980, Paul Benioff published a paper extending Turing machines, which are the mathematical ideal for classical computing (Chapter 2), to create a quantum mechanical version. This work shows that a properly designed quantum computer, like a classical computer, can carry out any possible set of logical or mathematical operations — just differently.

In Benioff's work, deterministic elements of the classical computing model are replaced by nondeterministic quantum versions. Just as Turing's work in the 1930s was vital to progress in classical computing, Benioff's paper laid the foundation for the current progress in quantum computing.

Speaking out on quantum computing

In 1981, the first Conference on the Physics of Computation was held at MIT. Benioff and Feynman gave talks on quantum computing. Benioff's talk was titled "Quantum mechanical Hamiltonian models of discrete processes that erase their own histories: application to Turing machines." It advanced his 1980 work showing that a computer can operate under the laws of quantum mechanics. Feynman proposed a basic model for a quantum computer, which he elaborated on soon after. Feynman famously declared, "Nature isn't classical, dammit, and if you want to make a simulation of nature, you'd better make it quantum mechanical."

TECHNICAL STUFF

Benioff's proposal focused on using quantum computers to replace or augment classical computers for certain operations, just as people are doing today, while Feynman's proposal focused on the use of quantum mechanical devices for simulating quantum mechanical interactions, which is also work that's in progress today.

Describing a universal quantum computer

In a 1985 paper, David Deutsch of Oxford University described a universal quantum computer — a quantum computer that, like a universal classical computer, could carry out any imaginable logical or mathematical operation. According to Deutsch's paper, it "would have many remarkable products not reproducible" by any classical Turing machine. The quantum gates in Deutsch's description operate in a way that's analogous to logic gates in classical computers.

Espying the CNOT gate

In 1986, Richard Feynman published a paper, "Quantum Mechanical Computers," describing an idealized quantum computer. Part of the purpose of the paper was to study potential physical limitations on computers of all kinds, including classical and quantum computers. The paper included a detailed description of a controlled not gate, called a CNOT gate, originally proposed by Charles Bennett and others in a paper published in 1973. CNOT gates (see Figure 5-3) are crucial to quantum computing.

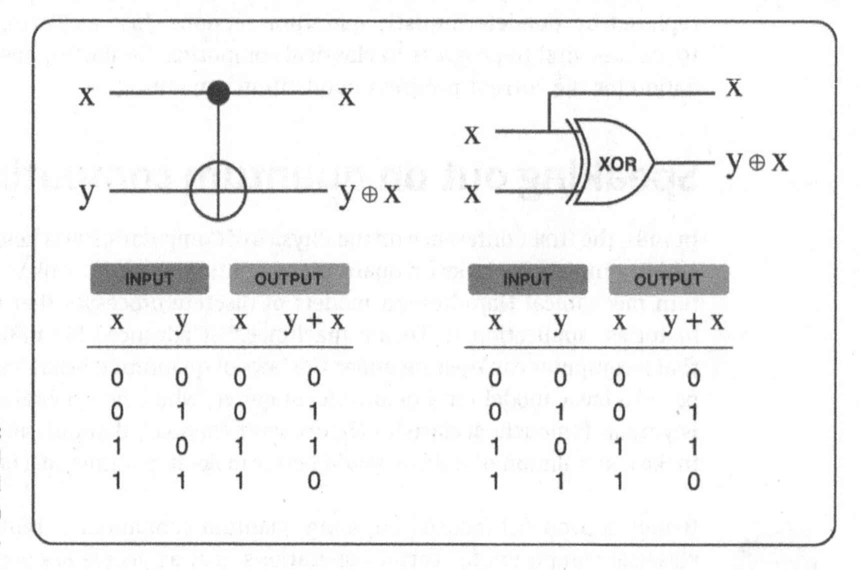

x	y	x	y + x
0	0	0	0
0	1	0	1
1	0	1	1
1	1	1	0

x	y	x	y + x
0	0	0	0
0	1	0	1
1	0	1	1
1	1	1	0

FIGURE 5-3: A CNOT gate, vital to gate-based quantum computing.

Modeling a real quantum computer

In 1988, building on previous work from Feynman and others, Yoshihisa Yamamoto and Kasuhiro Igeta created a proposal for a real-world, gate-model quantum computer, using atoms as qubits and photons for quantum communications. Their proposal included the CNOT gate. This proposal has served as the basis for current work in both areas.

A gate-model quantum computer features logic gates that are conceptually similar to the logic gates used in classical computers, but the gates are different.

Annealing for optimization

1989 saw the publication of a paper by Bikas K. Chakrabarti of the Saha Institute of Nuclear Physics in India describing the use of quantum annealing to solve optimization problems. Chakrabarti suggested that quantum annealing could help find better answers than could be deduced using principles from classical mechanics. This approach is now used by D-Wave Systems, the leader in quantum annealing systems.

Quantum annealing is part of an approach called *adiabatic quantum computation* — *adiabatic* meaning that the system does not gain or lose heat. This matters because the system begins in a lowest-energy ground state and stays there throughout computation, making it easier to protect the system from noise and to detect the influence of noise when interference does occur. Quantum annealing is potentially quite useful for optimization problems.

Breaking through in the 1990s with Algorithms and Hardware

The decade of the 1990s saw the intial explosion of interest in quantum computing, with three crucial developments:

- The initial realization that error correction for the state of individual qubits could be implemented effectively.
- The development of quantum-advantaged algorithms, especially Shor's algorithm, which threatens to make modern cryptography obsolete. (See Chapter 14.)
- The first working quantum computing hardware that could run an algorithm.

It was Shor's algorithm that really touched off widespread interest in quantum computing, while the arrival of working hardware added fuel to the fire.

Connecting with quantum communications

In 1991, Artur Ekert of Oxford proposed a method to achieve secure communication by using quantum entanglement. In 1993, an international group of scientists stated that quantum teleportation of information could be perfect, with both errors and eavesdropping being detectable by the intended sender and recipient. The only bad news: The process of accurately "reading" the source object scrambles its quantum state.

Figure 5-4 shows a visualization of quantum communication.

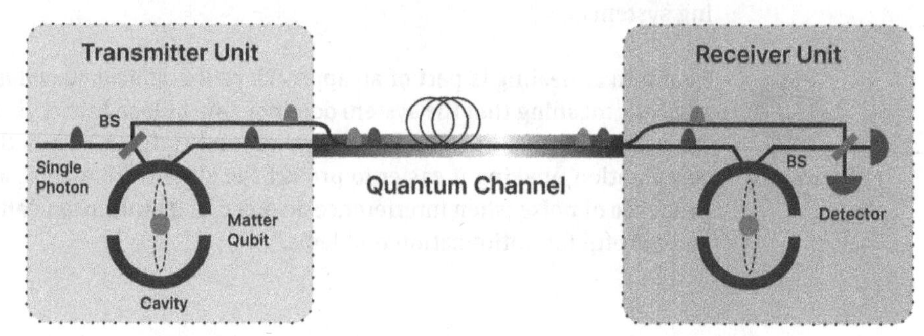

FIGURE 5-4: Quantum communication puts entanglement to work.

Unveiling the first quantum-specific algorithm

In 1992, David Deutsch and Richard Jozsa of Cambridge University created the Deutsch-Jozsa algorithm. (You saw Deutsch already earlier in this chapter — if this were a book on rope skipping, we'd call this double-Deutsch.) This new algorithm was the first quantum algorithm that could reliably be stated to be exponentially faster when executed on future, fast quantum hardware than the best possible classical mechanical version of the same algorithm when executed on a classical computer. This was the first mathematical proof of quantum superiority for an algorithm.

The Deutsch-Jozsa algorithm requires only a single function call to complete, although the problem it solves is a toy problem, meaning that the algorithm is not widely useful. The Deutsch-Jozsa algorithm was improved significantly by others in 1998 and, in 2003, was executed on an ion-trap quantum computer at the University of Innsbruck.

Shaking the multiverse with Shor's algorithm

It's rare that a mathematical algorithm, on publication, changes the world, but Shor's algorithm did. Among other things, it's highly likely that Shor's algorithm is the single most important factor (ha ha) leading us, decades later, to write this book — and a prime (groan) reason you're reading it.

Shor's algorithm is a quantum computing algorithm created by Peter Shor of Bell Labs. It's a much faster way to find the prime number factors of large numbers. This poses a problem for many of the encryption schemas used today to protect everything from credit card information to national security secrets to your email.

Usually, factoring large numbers is extremely time-consuming, making it very difficult to crack any code that requires a hacker to do so. Most encryption in use today takes advantage of this fact; the longer the key used, the more secure the code. The most secure versions of today's cryptography are designed to be safe from advances in classical computation for decades to come. Until Shor's algorithm appeared, there was no other kind of computation that could threaten this encryption.

So the arrival of a quantum leap in computing — pun intended — that might quickly render current cryptography useless was anywhere from concerning to terrifying, depending on what secrets you were depending on (which may be from business, the military, governments, or financial services organizations, among others). And since online banking depends on cryptography, and the world economy would quickly seize up if it were suddenly broken, the "terrifying" part is germane to all of us.

Current projections are that quantum computers that can break current cryptography are still many years away, but no one can be completely sure. So from 1994 to today, interest in quantum computing has skyrocketed, with Shor's algorithm as the single most important reason.

Shor, on a roll, proposed the first schemes for quantum error correction in 1995. We return to this increasingly important topic in Chapter 14.

Many private cryptographic keys are 128 bits, while others, considered more secure, are 256 bits. A typical public key is 2,048 bits in length. It's been proposed that these key lengths be immediately doubled to provide some protection against advances in quantum computing. Still, it's widely believed that using Shor's algorithm, quantum computers will break many commonly used public-key cryptographic protocols, including those based on the RSA (Rivest-Shamir-Adleman) and ECC (elliptic curve cryptography) approaches, in one or two decades.

Calling quantum for the defense

The US Department of Defense has been involved in a series of conferences on quantum computing, quantum cryptography, and related technologies. In 1995, Jonathan Dowling of Louisiana State University organized a particularly important international workshop in Arizona, which, as Dowling put it, "kick-started the entire Department of Defense program in these areas."

Militaries around the world are pursuing uses of quantum computing and related technologies, especially for communications. The work includes a mix of public and secret programs and publications. Figure 5-5 shows a recent visualization of potential defense uses of quantum technology.

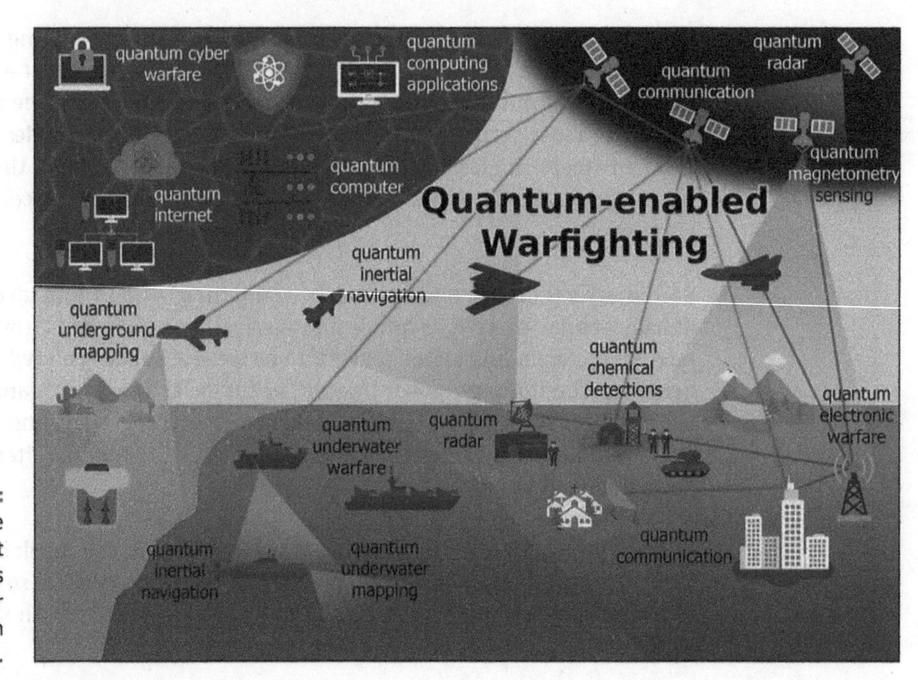

FIGURE 5-5: The US Department of Defense has big ideas for quantum technologies.

Initiating quantum error correction

Qubits tend to decohere during processing, so a quantum computer needs error detection and correction so it can restore an errant qubit and proceed. But the fact that quantum information can't be copied seemed to make any kind of protective backup and restore capability impossible. As a result, many researchers believed that useful quantum computing was impossible.

Peter Shor — yes, him again — along with Andrew Steane of Oxford, came to the rescue in 1995, proposing the first practical schemes for quantum error correction. Without copying the initial qubit, they showed how to spread its logical information across several mutually entangled qubits — nine, in Shor's work. Using previous efforts from 1985 by Asher Peres, Shor and Steane showed how to use this redundancy to detect errors and recover from them.

Just as Shor's algorithm was exciting tremendous interest in quantum computing — not least from the national security community, as evidenced by the DoD conference mentioned previously — Shor and Steane's error correction work showed that it might actually work, potentially even at scale. This good news struck the research community "like a bolt of lightning," said one participant. The change in opinion contributed to the first demonstration of qubit-based computing just a few years later and to progress since then.

Finding a groove with Grover's algorithm

Most important quantum computing algorithms promise immense speedups over classical computing alternatives. In 1996, Lov Grover of Bell Labs came up with a quantum algorithm that is not faster to quite the same degree as, for instance, Shor's algorithm, but still promises to save untold amounts of computing time when used, for instance, in search engines.

In classical computing, to find a match for one item out of *n*, you have to check items one at a time. You may find it on the first try, you may not find it until the last try, so on average it will take you *n*/2 tries. For instance, to search a list of 1,000 items, you'll need an average of 500 tries.

The notation used to describe the range of complexity of a problem is called Big O notation. For instance, a search algorithm that requires *n*/2 tries is still described as needing O(*n*), read as "on the order of n," tries. Similar imprecision applies to expressions such as O(n^2), read as "on the order of n^2," or O(sqrt(n)), read as "on the order of the square root of n." This shorthand is useful in providing a rough comparison between the expected difficulty of different kinds of problems for classical computers versus quantum computers.

Figure 5-6 is a visual representation of the process Grover's algorithm uses. Grover's algorithm cuts the number of searches needed to roughly sqrt(*n*), which is an exponentially smaller number. For instance, searching 1,000,000 items on a classical computer would take an average of 500,000 tries; searching 1,000,000 items on a quantum computer, using Grover's algorithm, would take roughly 1,000 tries, a reduction of more than 99 percent. (Many quantum algorithms promise speedups of far greater magnitude than even this.)

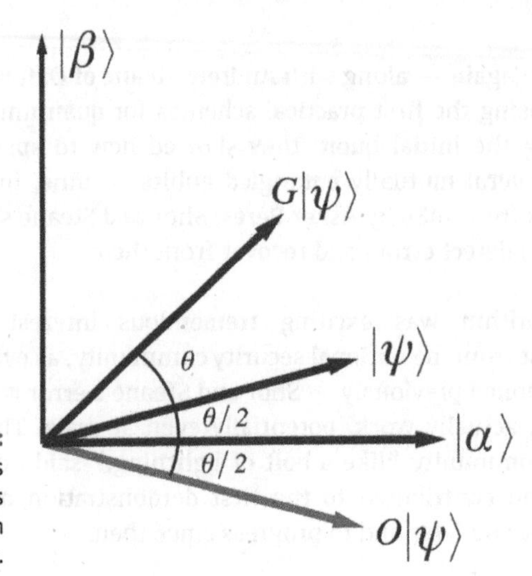

FIGURE 5-6: Grover's algorithm finds needles in haystacks faster.

One can only imagine the benefits of Grover's algorithm to, say, a search engine company. So it may be no accident that Google, of Google Search fame, is a big investor in quantum computer development. Microsoft, owner of the Bing search engine, is also working hard on quantum computing.

Manifesting a working quantum computer

Up until the late 1990s, most researchers believed that quantum computing was more a theoretical possibility than a potential reality. But Grover's quantum search algorithm did not sit on the shelf for long.

Late in the 1990s, scientists rather suddenly demonstrated solid progress on many of the problems that had prevented quantum computing from becoming real. The single biggest roadblock came directly from quantum mechanical theory: Trying to read the value of a superposition of quantum particles would collapse the superposition, without being able to determine what the final state had been. But by using nuclear magnetic resonance (NMR; see Chapter 4), scientists were able to align atoms into superposition and successfully read their state.

Today's quantum computers use a variety of techniques to induce superposition and manipulate the state of quantum particles without causing decoherence, as we describe in later chapters.

In 1998, physicists Jonathan A. Jones and Michele Mosca were able to run Deusch's algorithm on a 2-qubit NMR computer at Oxford University, as shown in Figure 5-7. Shortly afterwards, in the same year, physicists Isaac Chuang of IBM and Neal Gershenfeld of the MIT Media Lab and Mark Kubinic, a chemist from UC

Berkeley, successfully ran Grover's algorithm on a different 2-qubit NMR computer, loading the qubits with a small amount of data and outputting a solution.

FIGURE 5-7: The first working 2-qubit computer was created in 1998.

University of Oxford - Department of Physics

TECHNICAL STUFF

Both systems used NMR, and both were limited to a few nanoseconds of runtime before losing coherence. Many different qubit technologies are now in use (see Chapters 10 and 11), and runtimes, though longer, continue to be one of the largest challenges facing quantum computing.

Hitting seven lucky qubits in hardware

In 2000, quantum computing research continued to advance. However, scientists were still using various materials to perform computations; they weren't building machines for quantum computing anything like those we see today.

The highest qubit count up to that point was created by scientists at Los Alamos National Laboratory, who developed a 7-qubit quantum computer from a single drop of liquid — an acidic compound of simple hydrocarbon molecules.

In the same year, a 5-bit quantum computer was created at IBM Almaden Research Center. It used the nuclei of five fluorine atoms as qubits, programmed by radio frequency pulses and read by, as with previous devices, NMR instruments. In the same year, a textbook that's still standard, *Quantum Computation and Quantum Information* (Nielsen and Chuang, 2000) was published.

TECHNICAL STUFF

Note that the quantum computers in use at this time were conjured up one at a time from various raw materials. They were not a computer in the sense of being a standardized machine that could be reliably reproduced in the same configuration.

Starting Today's Quantum Computing Race

The accomplishments we list here are just selected highlights from a huge amount of work over a period of decades. It was only in the late 1990s that most researchers came to believe that working quantum computers were even possible. The creation of Shor's algorithm lit a fire under the field; demonstrations of actual quantum computation fed a great deal of fuel to that fire.

After everyone had finished partying like it was 1999 — because it was — the race was on. We describe the crucial accomplishments up to today in the next chapter.

» Putting more investment into companies

» Accelerating the pace of progress

» Investigating what's needed next

Chapter 6
Quantum Computing Accelerates

A lot of fascinating things happen in the world of science. You could cheerfully spend the rest of your life reading up on one scientific topic or another, such as gravity waves or the search for the Higgs boson, called the God particle. You would learn a lot and have a great time. But you might never see your newfound knowledge put to practical use in business, government, and the broader economy; many scientific projects never cross over to practical use.

Quantum computing is different; it's attracting tremendous interest and investment today and is likely to make a practical difference in the real world tomorrow. But why? What happened to make quantum computing different? And why did quantum computing take off? This chapter will answer those questions.

The developments described in Chapter 5 made quantum computing a focus of interest, investment, and excitement. But with these developments, quantum computing became more than just a promise; for everyone interested in computer security, including banks, governments, and national security agencies, it also became one of the biggest potential threats on the horizon.

Why? Because Shor's algorithm provides a potentially powerful way for a quantum computer to crack RSA encryption, the most-used encryption standard worldwide. When the algorithm was developed in 1994, no one had demonstrated quantum computing; it was a theoretical prospect. However, successful laboratory demonstrations of working qubits in 1998 brought this possibility far closer to reality.

Since then, national security people around the world have been waking up in the middle of the night in a cold sweat. Whereas children may go to bed on Christmas Eve with visions of sugarplums dancing in their heads, these national security people are having nightmarish visions of *Q-day*, which is the day quantum computers achieve enough qubits to "break the internet" by cracking computer encryption.

As mentioned in Chapter 5, today's encryption schemes depend on the extreme difficulty that classical computers have in factoring large numbers: something mature quantum computers, with perhaps thousands of qubits — many more qubits than we have today — are expected to be able to do.

Now we, your friendly authors, have a much more sanguine view of the situation. With regard to encryption, security measures have never been perfect and have always evolved. Adding more digits to current encryption keys could provide additional years of protection, while moving to quantum-based encryption may someday make secrets much more, not much less, secure.

Recently, in a step toward quantum-based encryption, several vendors, such as AWS, have launched services that utilize quantum cryptography methods to securely deliver code keys. This chapter describes this and other developments that have made quantum computing increasingly important — in research and development and in the real world.

Pushing Technical Progress Forward, 2000–2010

The late 1990s saw quantum computing become real. As the new century began, the research community kept adding fuel to the fire with new successes.

Seeing Shor's algorithm executed (Shor unharmed)

In 2001, researchers from IBM Almaden Research Center and Stanford created seven working qubits from among roughly a billion billion identical molecules, programmed by radio frequency pulses and read by NMR technology (similar to previous quantum computing lab work described in Chapter 5).

The degree of control that the researchers achieved over their qubits represented a milestone. They used these seven qubits and Shor's algorithm to factor the number — wait for it — 15. (Author Smith's 8-year-old granddaughter can do this too, and with much less trouble.)

Yet despite the simplicity of the task, it was still a landmark accomplishment. It lived up to the claim made at the time by Nabil Amer of IBM: "This result reinforces the growing realization that quantum computers may someday be able to solve problems that are so complex that even the most powerful supercomputers working for millions of years can't calculate the answers."

This device, and all the devices used in quantum computing work up to this point, were one-off lab creations that used qubits created in a test tube, not machines that might be able to be manufactured and used by people doing work. This is why the availability of working machines from D-Wave (2009) and IBM (2016), both described shortly, are so vitally important.

Making a quantum leap in education

Research in Motion (RIM) created the BlackBerry — an early and, for about a decade straddling the turn of the century, wildly successful smartphone. In 2002, RIM co-founder Mike Lazaridis funded and cofounded the Institute for Quantum Computing (ICQ) at the University of Waterloo, a visionary step that continues to resonate today.

SHOR-LY YOU JEST, MR. FEYNMAN

Richard Feynman gets much credit for his work in quantum mechanics and in other fields and for coming up with the idea of quantum computing, as well he should. But Peter Shor has done as much as anyone to bring Feynman's vision to life.

Shor won a share of the Breakthrough Prize in Fundamental Physics in 2022 and continues to toil in the vineyards of the field, making further contributions and teaching new generations of students at MIT. If you want to hear from him directly, you can view his video describing Shor's algorithm or follow him on Twitter at @PeterShor1.

Lazaridis has donated more than $100 million, ensuring that the BlackBerry isn't the only thing he'll be remembered for. Today, education in quantum computing and related fields is exploding, not least in Canada, with Lazaridis and the University of Waterloo among those leading the pack. The Institute for Quantum Computing, shown in Figure 6-1, continues to be a leader in quantum computing research and education today.

FIGURE 6-1: The Institute for Quantum Computing at the University of Waterloo.

University of Waterloo

A rich international heritage exists among quantum computing researchers. Among leading figures at IQC are the Canadian Roy LaFlamme, whose thesis advisor was the American physicist Stephen Hawking, and Italy-born Michele Mosca, advised by innovator Artur Ekert of Poland (see Chapter 5), whose thesis advisor was the acutely British Sir Peter Knight, a leading figure in quantum computing innovation in the UK. Artur Ekert's Scottish ties (don't ask) qualified him for a crucial fellowship during his education. There ought to be a book about this . . .

Experiencing spooky action at astronomical distances

In 2004 and 2005, several advancements were made concerning the use of entanglement between qubits. In 2004, five photons acting as qubits were entangled.

Five is the minimum number of qubits needed for quantum error correction, which requires entanglement of multiple qubits to create a single, working, error-corrected logical qubit. (Shor initially recommended nine entangled qubits to support each working qubit, but still.)

In 2005, scientists at the University of Illinois at Urbana-Champaign created qubits that used multiple characteristics of a single particle, raising the possibility that multiple qubits could be created per particle in future quantum computers. (That capability is not used in current quantum computers, but it may be in future ones.)

In the same year, the capability to measure the state of quantum bits without causing decoherence was demonstrated for the first time. (A capability that is not as reliably available for today's quantum computers as we would like; reading the state of qubits requires a restart.)

TECHNICAL STUFF

Quantum teleportation — the instantaneous transfer of a qubit's state, not the physical qubit itself — has been demonstrated in space. In 2017, Chinese researchers teleported the quantum state of a photon to a satellite in near Earth orbit, creating the first satellite-to-ground quantum communications network, as shown in Figure 6-2. The newly created network was immediately used to support a teleconference between Beijing and Austria. (To us, this was a "ground control to Major Tom" moment that the late, great David Bowie might have appreciated.)

FIGURE 6-2: Entanglement for quantum communications reached space in 2017.

JIAN-WEI PAN

Moving toward a fully implemented computer

Between 2006 and 2009, several steps were taken toward a fully implemented quantum computer. In 2006, the University of Waterloo and MIT teamed up to create the first 12-qubit quantum computer, again working with molecules in liquids to create qubits and using NMR technology to read the results.

In 2007, quantum mechanical versions of the transistor, RAM, and the data bus were created, raising the possibility of an entirely quantum mechanical computer. However, in today's quantum computers, only the qubits and their control circuitry demonstrate Quantum Technology 2.0 capabilities such as entanglement and superposition. It's easier and cheaper to use classical computers for the non-qubit parts of the computing machinery.

In 2009, the first experimental device with many characteristics of today's quantum computers was demonstrated. A team at the US National Institute of Standards (NIST) built a two-qubit computer using a logic-gate model — the first universal, programmable quantum computer. And in the same year, the US government put out its first quantum computing report.

Gate-based quantum computers are universal in the same sense as classical computers: They can perform any possible computing operation. The reason for pursuing quantum computing is its huge expected speed advantage for certain kinds of operations, once more qubits become available.

The logic gates in the NIST computer were encoded into laser pulses that manipulated individual qubits based on the spin of beryllium ions, with another laser used to read the results of calculations. In a large advance over previous experiments, which used molecules floating in a test tube, the ions were placed in traps within a gold-patterned aluminum wafer, a setup that will bring tears of recognition to the eyes of anyone who knows about modern microprocessor manufacturing.

The more modern setup was a large step in the direction of manufacturability for quantum computers. Ion-based qubits are in use today by several companies (see Chapters 10 and 11), of which IonQ is the best known.

Inconsistencies in the laser pulses caused an error rate of about 9 percent per qubit, resulting in an accuracy (called *fidelity*) of 91 percent (that's "one nine" of accuracy, in quality-control parlance). However, progress since then has resulted in fidelities as high as 99.8 percent. Many experts believe that accuracy of 99.5 to 99.9 percent or greater will be needed for practical quantum computing.

Also in 2009, Dorit Aharonov and Michael Ben-Or, building on work by Shor, showed that a quantum computer with a low enough hardware error rate — the figure they gave was 1 percent — can use error correction to further reduce the corrected error rate to nearly zero. This result was another "go" signal to large companies and other interested parties that had been tracking work in quantum computing to invest and get more involved.

Investing More Resources from 2010-2015

Beginning after 2010, business activity started to carry the torch forward. While energetic academic research continued, business accomplishments and new investment began to set the pace.

Shipping a working annealer from D-Wave

In 2011, in an act of optimism bordering on hubris, Canadian company D-Wave shipped the first commercially available quantum computer, the D-Wave One. This was a truly landmark moment in the history of quantum computing.

The announcement generated criticism at the time because D-Wave's machine is a quantum annealer rather than a quantum universal computer. Quantum annealers handle only optimization problems, not more general problems. However, many such problems exist, with a lot of potential value in solving them. (For example, oft-repeated processing loops in machine learning could potentially be sped up dramatically in this way.)

D-Wave has recently announced that it will use the lessons learned from shipping annealers to create an additional product line of logic-gate quantum computers.

TECHNICAL STUFF

You may be confused if you hear that quantum computers are limited because they have only a few hundred qubits, and then find out that D-Wave has been offering access to a machine with more than five thousand qubits since the year 2019. Simply put, the qubits in an annealer are not as capable as the qubits in a universal (logic-gate) quantum computer. So you have to consider qubit counts for annealers and universal quantum computers separately. (And yes, this is indeed confusing.)

Adding software to the incorporation party

In 2012, Canadian company 1QB Information Technologies (1QBit) was founded. 1QBit was the first quantum computing software company. Today, they have partnerships with many of the leading companies in quantum computing and in areas such as pharmaceuticals, financial services, online commerce, and consulting.

Since 2010, there has been a significant increase in deals and funding for quantum computing, with many companies and organizations investing in the field. While some funding is publicly announced, many large investments by organizations and governments are kept confidential. (See the Snowden comments in the next section.)

Causing a quantum panic with Edward Snowden

In 2013, Edward Snowden, a former intelligence officer, leaked a trove of documents containing US government secrets. Among the secrets exposed was the Penetrating Hard Targets project, an effort by the US National Security Agency (NSA) to develop a quantum-computing-based capability to crack cryptography. (This is the scenario we mentioned at the beginning of the chapter.) The exposure of these formerly secret NSA efforts confirmed suspicions that cracking cryptography with quantum computing was taken seriously by at least one national government. (Since spies are often called *spooks*, this added a whole new meaning to Einstein's famous description of quantum entanglement: "spooky action at a distance.")

The revelations from Snowden are also rumored to have inspired the beginning of large investments in quantum computing, including for similar purposes, by the Chinese government. Ironically, the US government, whose once-secret efforts inspired others, was criticized for waiting until several years later to publicly announce a similar, broad-based effort of its own.

Putting down (square?) roots in quantum computing with Google

Google started working on quantum computing in 2006 and went on to create the Google AI Quantum Team. In 2014, Google hired researcher John Martinis shown in Figure 6-3, and his team, from the University of California, Santa Barbara as part of an effort to build their own quantum computer. The company provides access to their systems through their Quantum Computing Service.

At this writing, Google has been relatively quiet about their quantum computing efforts for more than a year. We can't say if this is a temporary lull or a sign of a lessening of interest in quantum computing on Google's part.

FIGURE 6-3: John Martinis, upper right, worked at NIST-Boulder in the 1990s.

NIST

Bending the Arc of Progress Upward, 2016 to Today

In recent years, quantum computing has become widely accessible through the cloud, and something like a Moore's Law-type acceleration is starting to appear for superconducting qubit counts. Quantum computing is on the verge of showing dependable superiority over classical computers for a few algorithms, and business interest is increasing.

Offering quantum computing through the cloud

In 2016, IBM announced the IBM Quantum Experience — an interface to IBM quantum computing systems via the cloud — and changed how people access quantum computing.

Now most quantum computer manufacturers offer direct access to systems via the cloud. There are also three multiprovider offerings:

- Amazon Braket integrates with AWS services such as Simple Storage Service (S3).
- Azure Quantum features Microsoft software offerings such as the Q# quantum computer programming language.
- Strangeworks offers the widest range of systems.

Cloud offerings tend to include low-priced options and credits for those getting started. Cloud access to computing is described in more detail in Chapter 15. Quantum computing is now the most democratically available computing revolution in history.

Ruling that there ought to be a law

2018 saw passage of the US National Quantum Initiative (NQI), which sets out a 10-year plan for quantum computing and quantum communications applications. Among other provisions, the law established the Quantum Economic Development Consortium, funded through the National Institute of Standards and Technology (NIST) and made up of entities from industry, universities, and government.

The law provides a framework that can be enhanced and extended. For instance, in 2019, the White House issued Executive Order 13885 to create a National Quantum Initiative Advisory Committee with up to 26 members; this was expanded in 2022 with Executive Order 14073. Efforts under this law are likely to provide a useful underpinning for quantum information science in the US and beyond.

Governments are investing billions in quantum computing, not all publicly announced. The UK, for instance, recently announced a 10-year, $2B initiative. (There may be more of this particular iceberg below the waterline — that is, additional investment in secret projects.) International collaboration also appears to be strong, and hopes are that this will continue.

Going universal with IBM

In 2019 IBM announced the IBM Quantum System One, the first logic-gate-based (also known as universal) quantum computer to be offered for sale, with 27 superconducting qubits. IBM now claims to lead the world in quantum computing, and it's a hard claim to argue.

IBM's Quantum System One initially shipped with a 27-qubit processor. It can be upgraded to later processors with 65 qubits, 127 qubits, 433 qubits or — coming soon, at this writing — more than 1,000 qubits. This steady progress — so reminiscent of Moore's Law, as described in Chapter 2 — represents an amazing leap forward for quantum computing.

Both IBM's and Google's systems are based on superconducting qubits that are both programmed and read by a combination of laser beams and microwaves. Many other kinds of qubits are in active use; see Chapters 10 and 11.

Claiming quantum supremacy with Google and others

Later in 2019, Google claimed they had achieved *quantum supremacy:* the first reproducible example of a quantum computer quickly calculating a result that would have taken thousands of years to achieve on classical computers. This result was for a mathematically interesting but not practically useful algorithm running on their 55-qubit Sycamore machine.

IBM and others quickly claimed, in turn, that they had created a classical computer version of the same algorithm that approached the speed of Google's quantum result. The debate over this set of claims continues to this day.

Additional claims of supremacy have been made since. The claim most comparable to Google's was from a Chinese team running the same algorithm on a system with 66 superconducting qubits. The effort was led by Pan Jian Wei, widely referred to as the father of quantum computing in China.

Two similar claims have also been made on machines with photonic qubits, by a Chinese team in 2020 and by Canadian company Xanadu. While controversies continue, it seems clear that today's quantum computers can outperform classical computers on some tasks. Going forward, the question for current and potential users is finding where quantum computing offers advantages for their own computing challenges.

Getting entangled in a Nobel effort

In a moment that the quantum computing community and others around the world found both significant and moving, the 2022 Nobel Prize in Physics was awarded to three researchers who demonstrated quantum entanglement in increasingly restrictive experimental conditions.

The researchers tested Bell's inequality, an assertion made in an attempt to demonstrate that quantum entanglement does not exist. John Stewart Bell himself, who proposed Bell's theorem in 1964, eventually came to believe that entanglement is real, and the great majority of scientists today agree.

In three different decades, each of the three researchers made ever more detailed measurements under ever more exact testing conditions. (Although some of the experiments were conducted on small budgets and in the face of opposition or derision from parts of the scientific establishment.) In each experiment, the results supported the existence of entanglement.

We're also seeing entanglement put to use in the real world, both for quantum computing and for quantum communications. But the Nobel award recognized the critical achievement of experimental proof and the courage and dedication of scientists doing difficult work in conditions of great uncertainty — no reference to Heisenberg and his uncertainty principle intended — all over the world.

Asserting quantum utility with IBM

In mid-2023, IBM claimed to have achieved quantum utility. The claim has a fair amount of complexity to it, but it's worth the effort to understand at least a summary of what they achieved.

IBM calculated the interplay of 19 magnetic fields on both a classical computing supercomputer and a 127-qubit IBM quantum computer. For the simpler results, the supercomputer and the quantum computer produced nearly identical results. When the supercomputer ran out of gas, the IBM quantum computer was using 68 of its qubits.

The IBM quantum computer was able to continue up to the point where all 127 qubits were in use, modeling increasingly complex aspects of the problem. Since the supercomputer could not keep up, these most challenging computations were not considered verified. But they were in line with predictions derived from the results that were verified.

IBM was quick to acknowledge that future work on supercomputers may match or exceed what IBM's machine achieved. However, at the time the work was done, a quantum computer was the only way to achieve it. IBM calls this *quantum utility* — basically, quantum computing making itself useful for a real and complex problem.

Like every other quantum computer in use today, the IBM quantum computer did not have error correction capability. To get useful results, the IBM team used error mitigation techniques, including improvements to qubit control and advanced statistical techniques that used machine learning to identify likely errors and back them out of the reported results. Between substantive runs and error correction runs, the IBM team ran the simulation 600,000 times, which demonstrates the tremendous effort being expended on the cutting edge of quantum computing.

Many more results of this kind will need to be achieved before quantum computing can be considered a reliable part of the world's toolkit for solving the most complex computation problems. However, this result was a crucial step forward; we consider it to be on a par with the achievement of quantum supremacy, and it may in the future be considered to be even more important.

Finding Out What's Still Needed for Quantum Computing

Well, that was fast. (By *that*, we mean the quantum computing history described in Chapter 5 and this chapter.)

To sum things up all too quickly: Feynman had the original idea for quantum computers as quantum mechanical simulators in 1947. Research proceeded until exciting algorithms appeared in the early 1990s (Shor's and Grover's algorithms). Then, in a nearly insane rush of innovation, lab work created qubits in test tubes in the late 1990s and early 2000s.

The quantum computing switch had been pegged to "on." D-Wave was able to sell quantum annealers after 2010 and IBM introduced cloud access in 2016; at this writing, IBM is about to ship a universal quantum computer with more than 1,000 logic-gate qubits, and more than a dozen different providers are shipping systems. Investment and activity continue to increase, and steady progress continues.

Looking at the really big picture, it's roughly 100 years since the truly breakthrough ideas in quantum mechanics were put forward (Chapter 3), unleashing a flood of innovation from then until now. What's new in this wave of innovation — beginning with those first lab-created qubits, nearly 30 years ago — is the capability to create, maintain, and manage individual qubits with increasing effectiveness and skill. This wave of improvement in quantum computing capability has humanity on the brink of big breakthroughs in computing, communications, and measurement.

REMEMBER

Improvements in measurement using quantum mechanical sensors are a bigger deal than most of us realize. Just for one example, quantum-based improvements in the measurement of time and distance should make it possible to implement a local positioning system (LPS) with accuracy as good or better than today's global positioning system (GPS) devices, which require line of sight to satellites. (GPS access could be interrupted in, for instance, wartime.) Medical scans using new generations of quantum technologies may also offer great improvements.

Progress has been remarkable. Yet there is still a long way to go before quantum computing begins to approach the reliability of classical computing. When quantum computing does approach that point, amazing things will become not only possible but routine.

Three challenges jump out to us as authors and observers: Qubit counts need to increase; error correction needs to become standard; and the world needs greater quantum computing literacy to take advantage of all this power as it becomes available.

Qubit counts and error correction are interrelated. If error correction depends on virtual qubits, several physical qubits — we hope as few as 10 — will be combined into a single, error-corrected virtual qubit. Then qubit counts will need to zoom upward to create increasingly useful quantum computers.

Alternatively, with better qubit quality, the error correction "tax" may be less. And the tradeoff may be different for different types of qubits, as described in Chapters 10 and 11.

Error mitigation — different schemes to reduce but not eliminate errors — will precede and complement full error correction. Some algorithms and some use cases will find value with less than perfect error correction; others will require near-perfect or fully perfect results.

At the same time, user education is needed so that people in various fields are ready to take advantage of quantum computing as it matures. The faster quantum computing literacy spreads, the better. One of the authors (whurley) coauthored *Quantum Computing For Babies* (Sourcebooks Explore, 2018) with this purpose in mind, and now the two of us have teamed up to write this book. Educational efforts are exploding (see Chapter 16). It's been said that "it takes a village to raise a child." In a similar spirit, it will take efforts from all over the global village to raise the children who will take quantum computing to the next level.

2
Quantum Computing Options

IN THIS PART . . .

Identify where classical computing runs into limits and quantum computing excels so you can choose the right technology for specific types of problems.

Come with us for an introduction to quantum-inspired computing, which runs new kinds of programming algorithms on high-performance computers of the classical, not quantum, type.

Find out how quantum annealing offers a simpler but easier-to-build type of quantum computer that is being used today to optimize processes, materials, and objects.

Take your first step into the world of universally capable gate-based quantum computers, which are still in the early stages of development but showing increasing promise.

Come with us into the second step of gate-based quantum computers to learn how different types of qubits offer distinct advantages for future growth toward fully useful quantum computers.

» Seeing Moore's Law fall short

» Spotting promise in quantum computing

Chapter 7
Choosing Between Classical and Quantum Computing

Going forward, choosing when to use classical computing and when to use quantum computing will be a major consideration for all kinds of problem-solving. The advantages of using classical computers are pretty obvious, and classical computers are cheap and easily available, both as systems you buy and set up yourself and via cloud access.

To help you understand when you might need to consider quantum computing as a solution, in this chapter we set out the problems that classical computers face. Then we describe the architecture of quantum computing systems, both to make clear some of the challenges and to orient you as to the steps you need to take to get the most out of quantum computing.

Identifying Limitations in Classical Computing

Here's a quick summary of key problems in classical computing, followed by some of the details and a comparison to quantum computing. The limitations can briefly be summarized as follows:

- **The sunset of Moore's law:** Microchip speed is not increasing the same way it used to; in terms of ready computational power, classical computing is slowly running out of headroom.
- **Problems with exponential growth:** Important kinds of problems grind to a halt on classical computers, as described in Chapter 13. You'll need to look at quantum computing to see if it can help.
- **Fixed ways of thinking:** Your use of classical computing can get stuck in a rut. The emergence of quantum computing is inspiring new approaches in classical computing as well.

Watching the sun go down on Moore

Moore's Law, which we describe in Chapter 3, has been a useful guide for more than 50 years since it was put forth by Intel's Gordon Moore in the mid-1960s. In brief, it states that the number of transistors on an integrated circuit (as described in Chapter 4) doubles every two years.

Each doubling in the number of transistors per wafer has important implications:

- Feature sizes — which dictate the data bus lines and logic gates that make up an integrated circuit — get smaller.
- The speed of a given chip doubles.
- Costs for a given degree of capability drop by 50 percent.

In particular, microprocessors — the one-chip central processing units (CPUs) that are the brains of personal computers, smartphones, and many other devices — get twice as fast every two years, with little increase in price. Figure 7-1 shows a scatter plot of microprocessor speeds over time, depicting the rise in capability.

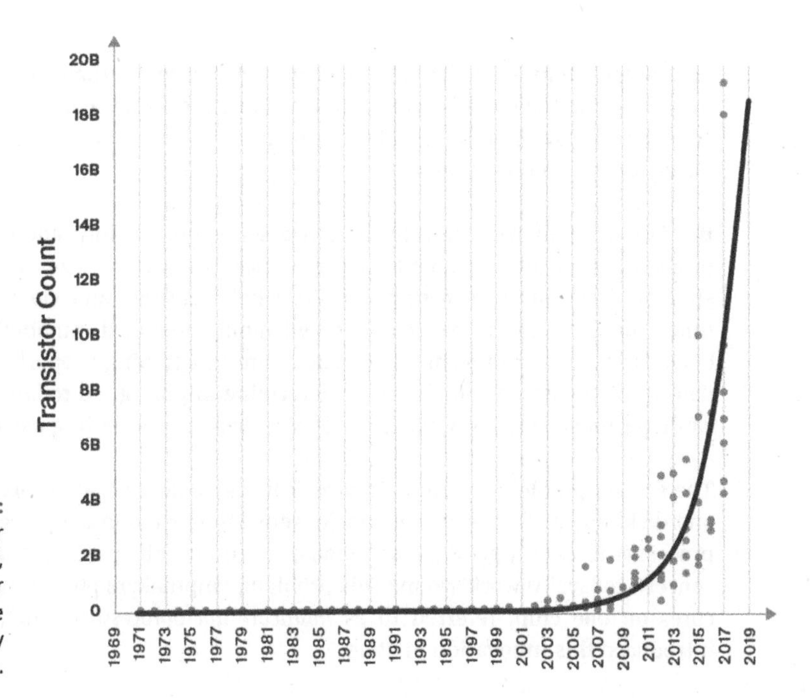

FIGURE 7-1: If you're 60 or younger, microprocessor speeds have doubled every year of your life.

Making devices with ever-smaller feature sizes takes ever-more-advanced technology. Taiwan is home to the world's leading company for this technology, Taiwan Semiconductor Manufacturing Company (TSMC). At present, even Intel, long a leader in device fabrication, makes its most advanced chips through TSMC.

Note that the vertical axis of Figure 7-1 is a logarithmic scale, so the actual improvement is huge. The rapid increase in chip capability is like getting a BOGO ("buy one get one" free offer) for everything at the grocery store for the next two years; then "buy one get four" for two years; then, "buy one get eight" for two years, and so on, for a person's adult life.

If Moore's Law had gone into effect at the grocery store starting in the early 1960s, the same dollar that got you a carton of a dozen eggs then would get you half a million such cartons today.

The compounding speedup provided by Moore's Law has been critical to computing and, indeed, to our modern era. IBM-compatible PCs needed processors that doubled in speed every two years to first run Word and Excel, then to move from MS-DOS to Windows, and then to run web browsers efficiently. And everyone has benefited as personal computers have rapidly dropped in price — they used to cost around $3,000 to $5,000, back when that was real money — and shrank in size and weight.

Your smartphone — which has more processing power than the computers that took people to the moon — can fit in your pocket only because of Moore's Law. (It could be even smaller, but most of us want big screens and big batteries, which is hardly Gordon Moore's fault.)

But Moore's Law has run into three problems, one obvious and two subtle. One problem, ironically, is quantum mechanics. As feature sizes get smaller and smaller, electrons stop acting like little ball bearings and start showing their inner, wavy nature. In particular, they tunnel, which can randomly switch a bit between 0 and 1 or prevent a logic gate from functioning properly. An inaccurate deterministic device can't actually determine anything, so today's feature sizes can't get much smaller without rendering classical computing useless.

The second problem is subtle. Devices can move data between processors, RAM, and disk only so fast. It doesn't make sense to put too much power into a single processor if it can't get data fast enough to run at full speed, and that is indeed a concern today. To work around this problem, chipmakers put multiple processing cores on one chip, referred to as *multicore microprocessors.* Intel, for example, introduced this innovation in 2005.

However, multicore microprocessors are useful only to the extent that you can smoothly divide a computer's tasks across the processors. Dividing problems up in this way is inherently inefficient, and the more cores you add, the more overhead and wastage of processing power you get. Before 2005, doubling the speed of a single-core processor would deliver the same improvement for every problem. Today, twice the cores doesn't mean twice the speed, and four times the cores even less so.

The third problem is even more insidious. The nature of computing has changed. PCs, smartphones, and other devices are increasingly used to communicate with the internet; the device you have your hands on is largely functioning as a smart terminal. Fast web servers do the data access and processing on the back end, and the right amount of computing power is automatically assigned to every job. So the bottleneck in performance moves from the microprocessor to the networking speed between your device and the cloud.

Speeding up the network is hard and expensive, and the typical connection is nowhere near doubling in speed every two years. Moore-type progress in processors matters less and less. In addition, ever-greater dependence on the internet opens up all sorts of connectivity and security problems, adding hassle factors and worse.

THE GOVERNMENT IS HERE TO HELP

Governmental help regarding quantum computing is available in most places. Many regions have "joined-up strategies," as the Brits call them, for supporting innovation in quantum computing. The UK just launched a new $2B effort; in the US, `www.quantum.gov` is the place to look.

A lot of governmental support and enthusiasm can be found for innovation in general, and quantum computing is a hot topic these days. Find out what's available in the regions relevant to your business and see if you can find allies at the applicable levels of government.

TECHNICAL STUFF

The introduction of 5G cellphone service is helping improve connection speeds for part of the networking journey. However, the range of a 5G tower is 1 to 3 miles, instead of the 40 miles plus for previous generations such as 4G. So people don't get all the 5G towers they need, and sometimes your phone can't reach a 5G tower at all, resulting in spotty service. Obstacles such as buildings and trees can interfere with 5G signals, further reducing their range and effectiveness. For instance, one of the authors (Smith) works in an office in Sunnyvale, right in the center of Silicon Valley. Ironically, cell speeds in the immediate area are awful.

Networking is part of the problem, but it's also part of the solution, in interesting ways. First, networking gives you access to the cloud, which gives you access to far more computing power than ever before.

Cloud access is a huge benefit, as shown by the rapid increase in cloud provider revenues as more computing moves to the cloud. And you don't have to make investments up front; cloud services are pay-as-you-go. The flip side of pay-as-you-go, however, is that as you go, you pay. Figure 7-2 shows the rapid rise in cloud computing spending.

Many organizations are pushing back hard on ever-rising cloud costs and trying to find less expensive ways to do things. Moore's Law brought more computing power for free, but it's running out of steam. Cloud services make it easy to get more computing power, but it's far from free.

Quantum computing is benefitting hugely from the cloud. Cloud access to quantum computer systems makes it much easier to get started than it would be if organizations had to buy their own quantum computers first. But quantum computing costs money too, of course. So as usage increases, some organizations may start bringing quantum computing systems in-house.

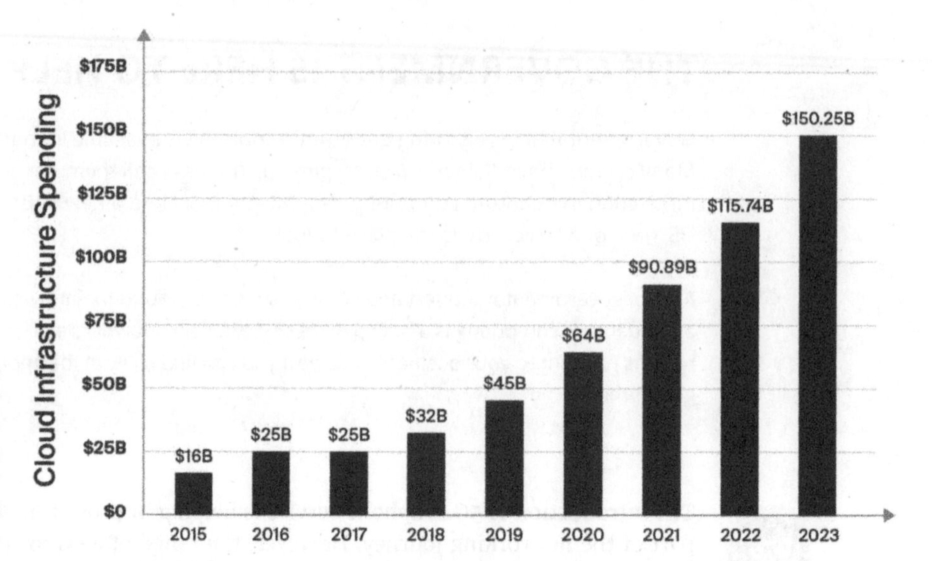

FIGURE 7-2: Cloud provider revenues — which also means cloud user costs — are rising very quickly.

WARNING

New technology can bring new challenges. 5G service uses more intense radiation than previous generations of cellphone service — and is able to direct that radiation toward individual devices (and their users). This exposure is said to not be a problem, but no one will have long-term experience with this until, well, the long term. In a similar way, responsible users of quantum computing need to be on the alert for potential problems arising from the applications they create.

All these problems interrelate to create a grim reality: Computing is no longer getting faster every two years, and not even close to that, when connectivity is taken into account. For specific problems, though, quantum computing may open up new vistas of performance.

Suffering exponential growing pains

Classical computing is great for most of today's routine computing tasks. You won't need quantum computing to post to Facebook or file your taxes anytime soon. But quantum computing is amazing at handling many computationally intensive problems. For instance, optimizing routes for a fleet of delivery drivers is likely to be a worthwhile use of quantum computing.

REMEMBER

An *algorithm* is just a set of steps followed by a computer, a person, or the two working together to solve a specific kind of problem. When you add two large numbers using pencil and paper (or in your head, if you're showing off) you use an algorithm to do that, although you may not be consciously aware of the steps you're following.

Algorithms are crucial in quantum computing because today's classical computers handle many tasks very well; it's only for specific algorithms, involving large numbers of data points, variables, or options, that quantum computers deliver much speedup.

However, for those algorithms, as the power of quantum computers grows, the speedup will range from worthwhile to incredible. Machine learning is an example of an area where quantum computing is on the cusp of making a big, big difference.

One prominent example of when classical computing slows down is when there's a *combinatorial explosion*, which is when a moderate increase in the number of inputs causes a huge increase in the number of calculations a program needs to make.

The easiest example to grasp is the traveling salesperson problem. Let's say you want to find the shortest route among 10 cities — or among 10 stops on a delivery driver's package delivery route. There are more than 180,000 combinations, which is manageable for a laptop computer. But for just 15 destinations, the number of possible routes exceeds 87 billion. That would take about half a day to calculate, depending on which algorithm and which computer you use.

Increasing the number of destinations to 20 ups the calculation time just a wee bit — to 2,000 years. Imagine if the classic movie *2001: A Space Odyssey* had a running time of 2,000 years instead of 2 hours. And imagine that the whole movie was spent watching HAL, the antihero computer in the movie, calculate the time required for different routes starting and ending at Jupiter? Not a fun way to generate a very, very large cloud-computing bill. (The only advantage is that you would be long gone by the time the bill came due.)

Many logistics problems have a lot more than 20 destinations, and they include many interactions and tradeoffs, not just routing for one salesperson. Delivery drivers for UPS and others face this problem every day. And trying to route a city's buses for a day means looking at the best route for combinations of buses. Then you want to take into account the number of drivers available that day, bus breakdowns, bad weather, and more.

Classical computers can solve problems like the traveling salesperson problem only in exponential-type runtime; the time needed rises hugely against small increases in the number of inputs (cities to visit or buses to route), such as the increase from 15 to 20 cities in the traveling salesperson problem. Quantum computers can run many of these same problems in polynomial runtime. Here's what that means:

- **Exponential-type runtime:** On a classical computer, the number of inputs becomes an exponent. For instance, a problem with *n* inputs may require 2^n calculations, a number which grows rapidly (see Figure 7-3); the traveling salesperson problem is a fairly severe example of this.

» **Polynomial-type runtime:** The number of inputs is the base, which may have an exponent or a multiplicative factor applied to it. For instance, a problem with n inputs may require n^2 calculations, which grows much more slowly (also shown in Figure 7-3).

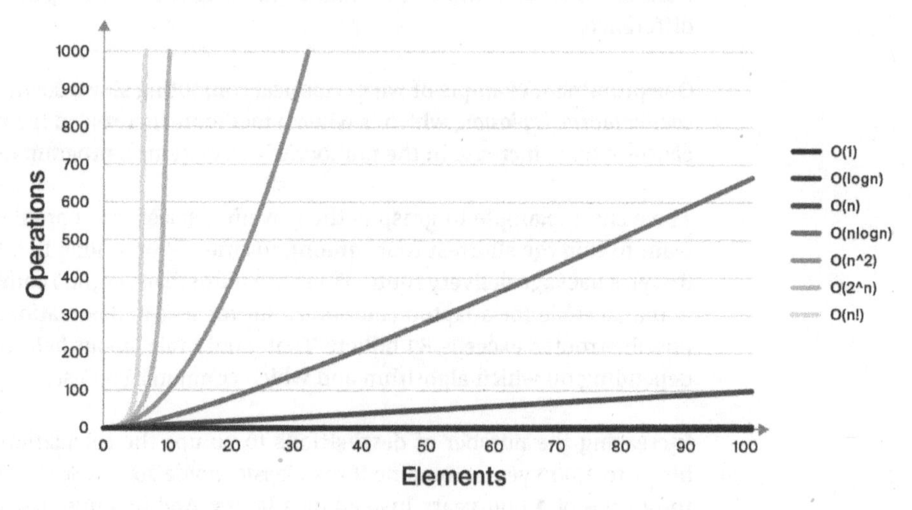

FIGURE 7-3: The number of calculations for exponential-type runtimes grows very quickly, while polynomial-type runtimes grow slowly and steadily.

Why are quantum computers faster? Let's say a truck has 20,000 possible routes, and it takes a classical computer 1 minute to figure out the amount of time each route would consume. To figure out the fastest route, the classical computer would need 20,000 minutes (about 2 weeks) to calculate the time needed for each of the routes, then a few seconds to compare the times and identify the lowest time (and therefore the fastest route). Waiting two weeks for results would hardly be sensible for dispatching delivery trucks on a daily basis.

On a quantum computer with enough qubits, the quantum computer could compare all 20,000 routes in a single program run, which might take only a few seconds. This approach would be very sensible for dispatching delivery trucks on a daily basis.

TECHNICAL STUFF

Want to know more about the difference between classical computing and quantum computing algorithms? See Chapter 14, where we introduce several major types of quantum computing algorithms, including the problems they solve, the speedup they offer, and any limitations on their use.

WARNING

Today's quantum computers have a limited numbers of qubits and no built-in error correction, which means it's difficult or impossible to get the exact answer for larger calculations. However, even with these limitations, you may well be able to get a very good (but not exact) answer much faster than with classical computing. For some problems, an inexact answer calculated rapidly is an improvement over what's available from classical computers today.

Fighting fixed ways of thinking

You would think that 100 years of work with classical computing would mean that a lot of approaches would be worked out to the seventh decimal point. "But noooooo," as the late, great John Belushi used to say. Surprisingly often, this is not the case; when quantum computing notches a new achievement, classical computing fans often find surprising new ways to make their systems work faster.

Let's look at an example where a lot of improvement has indeed taken place. Many database queries are written using structured query language (SQL), which was defined 50 years ago. SQL is supported very well by relational databases — generally, databases that store information in rows and columns. (A driver's license database is a simple example.)

Because SQL is widely used and has been around a long time, it's been optimized by academics, database companies, hobbyists — you name it. So it's very fast, given the amount of work it can do, and the capability of a database to support SQL efficiently is a major selling point.

However, many problems in computing turn out to not be optimized as well as they could have been. To illustrate this fact, in the brief history of quantum computing, the following sequence of events has already happened several times:

- Someone has a problem that runs slowly on a classical computer.
- Someone else with a quantum computer runs the problem much faster and claims quantum superiority.
- The original someone gets inspired and writes a new algorithm for the classical computer, often using quantum-type techniques that run on classical hardware, reducing or eliminating the difference.

We call the new, classical solution described in the final bullet *quantum inspired.* This sequence of events is often claimed as proof that we don't need quantum computing. However, it took the challenge from quantum computing to make conventional computing programmers sharpen their pencils and find new ways to look at the problem. And the improved classical computing results push the quantum computing people to greater efforts on their side as well.

This back-and-forth is helping people to get the best out of *both* the classical system *and* the new quantum computing systems. As mentioned in previous chapters, high-performance classical computers are now being designed to share problems smoothly with quantum computers. As quantum computers improve, the challenge they pose to classical computers will increase, again leading to better overall results.

The bottom line

A theory in economics states that people will use an existing product or service until they run into *failure to supply* — the provider stops giving the customer what they're paying for. Only then does the customer look for alternatives.

Classical computing has a failure-to-supply problem when it comes to increasing speed and power, the capability to scale up to truly large problems, and, in some areas, flexible thinking. Customers expect classical computing to keep improving rapidly, and in some cases, that is no longer happening.

Quantum computing is stepping up to fill the gap. Knowing where these gaps are will help you make good decisions about when to try a quantum computing solution.

BUT WHAT ABOUT ChatGPT?

AI and machine learning are improving rapidly. The emergence of ChatGPT and other tools developed by mining reams of data from the internet has awakened a lot of interest, investment, usage, and increasingly even real productivity, almost overnight.

One of the coauthors of this book (whurley) used ChatGPT in March 2023 to innovate at the South by Southwest conference in Austin, Texas, where he lives. whurley gave the conference keynote presentation, to warm applause. He then announced at the end that he had generated the entire talk that morning using ChatGPT, which he had primed with many of his existing writings. This was a first for "South by," as people call it, and perhaps for the world.

So, as the recent improvements in AI show, classical computing certainly still has some cards to play — but with some limitations. Running machine learning algorithms on large amounts of data often results in exponential-type runtimes. AI programs are often very computationally intensive, energy intensive, and expensive to run, while the results are still suboptimal.

One of the biggest opportunities for quantum computing is to incorporate it into the inner loops of machine learning algorithms, reducing their use of time, computing power, and energy. The ChatGPT-type programs and other AI programs of the future may far outperform today's efforts as quantum computing is integrated into the solution.

Finding What's Right with Quantum Computing

Now that we've shown the challenges with classical computing, several questions naturally arise:

- Where is quantum computing better?
- Who needs the things quantum computing is better at?
- What's the status of the relevant quantum computing capabilities today?
- Who gets early access to the best new quantum computers?

We take you through general answers here, and then dive into the details — of systems, software, services, and educational offerings — in the chapters that follow.

Because we're still at the beginning of the quantum computing race, plenty of room exists for imaginative new solutions. These will come from all sorts of places. Uber was started because a couple of friends couldn't get a taxi; Airbnb was started because one of the company's founders couldn't afford a hotel room for a conference he wanted to attend. Let your imagination soar when thinking through new applications for quantum computing.

Highlighting where quantum computing is better

Fortunately for human progress in the years ahead, quantum computing is good at the very thing that classical computers are bad at: problems that result in a combinatorial explosion, as described previously.

Problems of this type include the following (roughly in order of complexity):

- **Optimizing combinatorial problems:** The traveling salesperson problem and knapsack packing — putting the most items of different sizes into a fixed volume of empty space. Business applications include routing of many kinds, logistics problems, and portfolio optimization for investors.
- **Linear algebra problems:** Matrix math and principal components analysis useful for machine learning and many other tasks. Business and science tasks include risk management, marketing, and sequencing for DNA classification.

» **Factorization:** Breaking large numbers down into their prime factors as used in decryption. Business and science tasks include computer security and mathematical research.

» **Differential equations:** Modeling the behavior of complex systems such as fluid dynamics (see Figure 7-4) and quantum particles (such as the atoms in complex molecules). Business and science tasks include vehicle design (cars, planes, boats), medicine (making a better pacemaker or stent), and molecular simulation (drug discovery and other medical research).

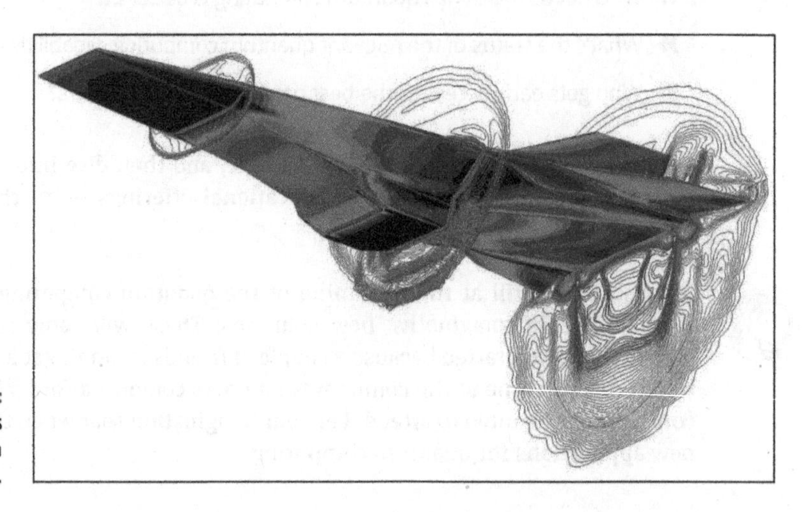

FIGURE 7-4: Fluid dynamics are crucial in airplane design.

REMEMBER

A *combinatorial explosion* describes a problem in which small increases in the number of inputs cause a large (usually exponential) increase in the processing time. A spectacular example is the traveling salesperson problem, where small increases in the number of cities visited leads to immense increases in processing time; future quantum computers with thousands of qubits will be able to handle this problem far faster. As mentioned, a low-end example is search, where a classical computer needs to examine an average of 1/2*n* items to find an answer, with *n* being the number of items in the list being searched. A quantum computer needs only sqrt(*n*) searches, which can be a very large saving indeed, though still less than can be achieved with the traveling salesperson problem.

To get the benefit of quantum computing, you have to find or create an algorithm that both exhibits superior performance on a quantum computer and solves your problem — or you have to restate the problem in a way that meets these conditions. If you can evaluate and play with algorithms, you'll have a great deal of fun with quantum computing; if not, you'll need to find people who can help you navigate the options.

WARNING

You may not be able to find an algorithm that helps with your problem. Or you may find the algorithm, but using it for the problem you're trying to solve may not be possible with the low qubit counts and lack of error correction in today's quantum computers. As the British say, "suck it and see."

Table 7-1 shows a few examples of early tests, dating back a decade or two, of the most powerful and influential current quantum computing algorithms. The first two are mentioned in Chapter 5, and the third in Chapter 6. Only the quantum annealing algorithm can be used for solving practical problems on today's quantum computers.

TABLE 7-1 Proof-of-Concept Implementations of Quantum Algorithms

Algorithm	Technology	Problem Solved
Shor's algorithm	Bulk optics (2012)	Factorization of 21
Grover's algorithm	NMR (2000)	Unstructured search, *N*=8
Quantum annealing	D-Wave 2X (2015)	Ising model on a Chimera graph with 1,097 vertices
HHL algorithm	Bulk optics (2013 and 2014), NMR (2014)	2×2 system of linear equations

Courtesy of Nature

Here's a brief description of the use of each algorithm:

- **Shor's algorithm:** Factors numbers; useful in breaking specific codes
- **Grover's algorithm:** Speeds search through lists
- **Quantum annealing:** Provides optimized solutions under constraints
- **HHL algorithm:** Provides approximate answers to systems of linear equations

TECHNICAL STUFF

People who are good at math are good at restating problems in different terms so they can be solved with more powerful or easier-to-use approaches. For one example, in the development of quantum computing, Walter Heisenberg, Max Born, Pascual Jordan, and others pioneered the use of a then-obscure math approach called matrix mechanics for use in quantum mechanics, introducing this in 1925 to the applause of almost no one. But a year later, Erwin Schrödinger introduced wave mechanics, which does the same thing in a completely different way. Schrödinger's approach was already familiar to physicists and was widely adopted, leading to much greater popularity and use for the new set of theories. Similarly, quantum annealing can be used with Ising models or quadratic unconstrained binary optimization (QUBO) formulations, but these are mathematically equivalent and easy for those in the know to convert back and forth.

Querying quantum's status today

Quantum computing technology is improving quickly. "We don't need this" can turn to "We've got to have this!" quickly. So we suggest that you follow Wayne Gretzky's advice to skate to where the puck is going to be — that is, start learning about quantum computing now. You've started to do that just by reading this book.

But once you've laced up your skates and glided forth on your quantum computing journey, the current state of play becomes an important navigational aid. Today's quantum systems are quite limited.

Today's quantum computers may be faster than the classical computer simulation for a given problem, but not by light-years. And there are some issues on the quantum computing side:

- Today's quantum computers offer raw, un-error-corrected physical qubits, which means real difficulty in getting results you can have confidence in.
- Quantum computing requires a big learning curve. You've started on the curve by reading this book, but a lot of learning is still in front of you.
- There aren't many teammates yet. The very definition of being an early adopter is that your entire QC team may be just you.

So why get involved? Because now is the time to gain and sharpen skills in quantum computing. And the same goes for organizations as a whole. You can learn where you can get an edge today and be ready as the picture improves. Within just a few years, qubit counts will be better, error rates will improve, and you will be among the first to gain a real advantage.

Looking for early access

In all areas of technology, there's a pecking order as to who gets access to the latest and greatest systems. In quantum computing, this pecking order is really important.

We are at a point in the development of the technology where it may go from interesting to absolutely indispensable in a single technology release. And this kind of lightning is likely to strike differently for different algorithms, application areas, and use cases. In fact, only the relatively few people who are already plugged in even feel the tingle of electricity in the air.

IS THERE A QUANTUM COMPUTING MOORE'S LAW?

Efforts in the quantum computing world appear to be trying to emulate Moore's Law–type progress in the quantum realm. Take a look at IBM's roadmap, as shown in Chapter 1. It looks a lot like Intel microprocessor roadmaps from 30 years ago (kindly don't remind us that that's before some of you were born).

Now even IBM's roadmap puts us a long way from the million-qubit systems that some experts say may be necessary for a system to truly fly in the era of quantum computing, given the apparent need to use much of the qubit budget for error correction. (Although doubling is incredibly powerful; if you start with 400 qubits or so at this writing, that number has to double only 12 times to get to roughly 1.6 million.)

In addition, qubit counts are not enough on their own. Those qubits need to maintain coherence (usefulness) long enough to run complex algorithms, and error correction is a challenge likely to impede progress until it's fully solved.

REMEMBER

The author William Arthur Ward said "The pessimist complains about the wind; the optimist expects it to change; the realist adjusts the sails." When it comes to quantum computing, it's easy to focus on all the challenges facing the industry and take a wait-and-see attitude. However, quantum computing is likely to hit an inflection point that will begin a period of very rapid change in computing. In fact, that change may already be underway by the time you read this. We suggest that the realistic thing to do, if you're in IT or will be affected by quantum computing, is to adjust the sails of your career and plans today.

So here's the rough pecking order for quantum computing access:

- **Quantum computer manufacturers:** Building your own quantum computer means never having to hear "I'm sorry" when it comes to getting access. IBM, Google, and quantum computing vendors such as D-Wave and IonQ always have first dibs on the latest and greatest systems their company can create.
- **Cloud platform companies:** Platforms such as AWS, Azure, and Strangeworks get early access to devices from the manufacturers who make their systems available using the cloud platform company in question, because they're such an important channel for reaching the market.
- **Consultancies and application vendors:** Consultancies advise their customers on future directions and application vendors need lead time for future software releases. So providing early access to these independent players is in everyone's interest.

» **The rest of us:** Quantum computing supply is ahead of demand at this point, so access to most systems is provided widely once a system is released. Credits for free use are also available for new users and researchers. However, getting access to the very latest and greatest systems is difficult.

TIP

Contributing actively to relevant open-source projects or interest groups can be a great way to gain insider knowledge and influence in various technology fields. This approach can be particularly beneficial for those who are not particularly technical and want to learn more about a specific area. The key is to focus on having a positive effect and helping others without being (solely or visibly) focused on gaining insider knowledge and influence.

If we can speak plainly, it's good for you to make friends in quantum computing now. ("You" means yourself as an individual and also any organizations that you're a part of or lead.) And the best way to make a friend is to be a friend. Find out who's doing cool things in the areas of interest to you and see how you can help them make progress. Attend conferences, listen to pitches, and get oriented.

We believe there will be a previously unseen gold rush in quantum computing, sooner rather than later. And there will be smaller gold rushes in specific areas of interest as progress brings first one problem area, and then another, into range of the new and improving quantum computing artillery. You want to orient yourself and establish alliances before these surges in interest, not when everyone is breathlessly rushing toward a Next Big Thing.

Needing what quantum computing offers

Everyone's interested in a better mousetrap, but you only need it if you have mice. So at this point, you have to explore whether you have, or expect to have, the mice that the quantum computing mousetrap is going to catch.

You also have to examine whether you need to buy, master, and use the mousetrap yourself. For example, if you have a robust in-house AI development effort, future advances in machine learning powered by quantum computing will be indispensable. But if you're using AI models provided by others, future speedups generated by quantum computing speedups may be provided to you by vendors. Figure out in advance what work you'll have to do yourself and what work you can rely on suppliers for. And then turn to Chapters 13 and 14 for more information.

» **Focusing on algorithms**

» **Involving your organization**

» **Looking into quantum-inspired solutions**

Chapter 8
Getting Started with Quantum Computing

The quantum computing ecosystem is more complicated than many people realize. To get value out of quantum computing today, you need to understand the main types of solutions and what each one is best for.

In this chapter we describe how to map different kinds of computing challenges to different kinds of quantum computing solutions that will unfold over time, so you can get off to a productive start with this emerging technology. Then, as quantum computing advances, you can get more and more value from it.

Then we introduce quantum-inspired solutions, a surprise contender as the near-term winner for useful quantum computing technology. Chapters 9, 10, and 11 walk you through the quantum computing hardware and software stack and the different kinds of quantum computing solutions available today. In later parts of the book, we show you how to get hands-on experience and start putting quantum computing to work for yourself and your organization, guided by the overview of available solutions in these chapters.

Identifying Five Classes of Solutions

Five classes of quantum computing solutions are available for you to consider — and the first two might surprise you:

- **Quantum-inspired classical computing:** Quantum-inspired classical computing means quantum-inspired computing algorithms running on classical computers. Cloud platforms — at this writing, AWS, Azure, and Strangeworks, along with competitors — and high-performance computing vendors are stepping up to provide quantum-inspired solutions. (Supercomputers are the best-known type of high-performance computer.) More on quantum-inspired solutions later in this chapter.
- **Quantum annealing:** Quantum annealers provide useful solutions to optimization problems, including some specific problems in machine learning. In some cases, those solutions are available and may give you an advantage today; in others, you should be able to get value within the next few years. You can access quantum annealers through the cloud via the leading vendor, D-Wave, through access vendors, or via cloud platforms. For more on quantum annealing, see Chapter 9.
- **Other analog quantum computers:** In addition to quantum annealers, there are additional types of analog quantum computers. These systems handle a wider range of problems than annealers, but they're currently less mature. And they still do not handle the full range of problems that universal quantum computers can handle. We don't talk about these systems much in this book, but they may well become more important in the next few years.
- **Universal quantum computing:** Practical solutions from universal quantum computers will address a wide range of problems but are still a few years away. However, if you have problems or opportunities that fit these solutions, you may want to get started now so you'll be ready when advantages for these platforms solidify. Universal quantum computing platforms are described in Chapters 10 and 11.
- **Quantum computing emulators (aka simulators):** A quantum computing emulator/simulator allows you to run quantum computing programs on classical computer hardware. They're great for learning quantum computing — even on your laptop — and for testing and building error control and other technologies for tomorrow's quantum computers. Most emulators/simulators map to universal quantum computers (see the preceding bullet), but you may encounter them for quantum annealers and other analog quantum computers as well. We mention emulators/simulators and their use in quantum computer programming in Chapter 16.

A quantum computer emulator allows a classical computer to mimic a quantum computer. However, many in the industry, including Google and IBM, use the term *quantum simulator* as the name for their quantum computer emulators. (We hereby decry the widespread deficiencies in liberal arts education that probably caused this mistake.) We will sometimes use the term *quantum computing emulator/simulator* to help you map the correct term, emulator, to simulator, the term that's become widely used in the industry. (An actual simulator — something that took the user inside a qubit, for instance — would be a blast.)

Why might this full taxonomy be surprising? Because today, and in fact since Shor's algorithm appeared 30 years ago, almost all the focus has gone to universal quantum computing. And that is the only type of quantum computing that will someday be able to run Shor's algorithm at a speed and level of effectiveness that threaten today's widely used cryptographic standards.

In the future — perhaps by the end of this decade — universal quantum computers may start to crowd out other kinds of quantum computers. Today, however, universal quantum computers are said to be in the noisy intermediate-scale quantum (NISQ) era. Here's what the NISQ for universal quantum computers means:

- **Noisy:** Error rates are high because there's currently no error correction and only the beginnings of error mitigation. For any given result, you don't know for sure whether it has been affected by errors.
- **Intermediate-scale:** A nice way of saying that these computers are only beginning to have enough qubits to be useful for practical problems — and again, these qubits are error-prone (see "noisy"), reducing their usefulness.
- **Quantum:** The term NISQ again uses *quantum* as if universal quantum computers were the only possible solution, leaving out quantum-inspired solutions, quantum annealers, and other analog quantum computers.

Error mitigation uses a variety of techniques to reduce errors; *error correction* is a more fundamental process that almost eliminates them. The better the mitigation you have in place, the less work there is for error correction to do.

REMEMBER

The *universal* in *universal quantum computers* means that a computer is a Turing machine — capable of solving literally any mathematical or logical problem, once said problem is reformulated into steps the universal computer can handle. Classical computers are also universal computers, but classical computers have an acceptably low error rate and universal quantum computers are not yet close to that. In the long run, classical and quantum computing will complement each other very well.

REMEMBER

Also remember that Turing's machine depends on an infinite tape (or other data store) and that nothing in our fallen world manages to be infinite, except perhaps for the power of love, which no one has yet manifested directly into a physically instantiated computer. So when we refer to either today's classical computers, or future, error-corrected quantum computers, as being *universal*, there's a *but* in there. And as so eloquently stated in the comedy classic *Pee Wee's Big Adventure* (1985), "That's a mighty big but."

Here's how each type of quantum computing solution addresses the error rate challenge:

- **Quantum-inspired:** This term is shorthand for quantum-inspired algorithms running on classical computers, a good approach for optimization problems. Classical computers have full error correction, so there's no problem with error rates. However, the speedup from quantum-inspired solutions is not nearly what universal quantum computers will deliver in the future.
- **Annealers and other analog quantum computers:** This term is true quantum computing but does not meet the "universal" criterion. Quantum annealers and their analog brethren are used for only optimization problems — such that an answer muddled by errors but delivered quickly is superior to an exact answer that takes a long time to reach on a classical computer. Fortunately, many machine learning processes can use optimization, so there's a lot of promise in this area.
- **Universal quantum computers:** Currently, error rates are still too high and available qubit counts are still too low for universal quantum computers to be useful for practical problems. As the error correction challenge is met and qubit counts increase, universal quantum computers will become more and more powerful.

Today, the alternatives to universal quantum computing solutions — quantum-inspired classical computing solutions and quantum annealing — are where many believe you currently can, or soon will be able to, find real business value. So they deserve your attention.

And when will these solutions be available? We can generally class their availability, as of this writing, as being today, tomorrow, or the day after tomorrow:

- **Today:** Cloud platforms and high-performance computing vendors are rushing to develop and deliver quantum computing emulators/simulators and quantum-inspired classical computing solutions to market. D-Wave is aggressively developing and delivering quantum annealers. Others are doing early work on other kinds of analog quantum computers. Universal quantum solutions for a wide range of problems are moving from research to development.
- **Tomorrow (before 2030):** Quantum-inspired and quantum annealing options for machine learning will continue to mature and begin to deliver real business value for optimization problems and specific optimization-type machine learning problems. Other kinds of analog quantum computers might get traction. Universal quantum computing solutions will start to deliver systems at greater scale and begin to deliver business value for specific problems.
- **After tomorrow (2030 and beyond):** Quantum-inspired and quantum annealing systems become more robust and more widely useful, increasingly joined by other analog quantum computing options. Universal quantum computing solutions begin to deliver business value for a wide range of problems, changing the world in the process.

The current situation as to usefulness is summed up in Figure 8-1. The picture is currently the same for quantum-inspired solutions and quantum annealing, but this may change over time.

We kindly request that you accept this framework as a way of looking at the future, not as a solid set of predictions from your humble authors. And we can only hope the quantum computing future does not end up resembling the 2004 movie *The Day after Tomorrow*, a climate change disaster movie in which the Northern Hemisphere largely froze solid.

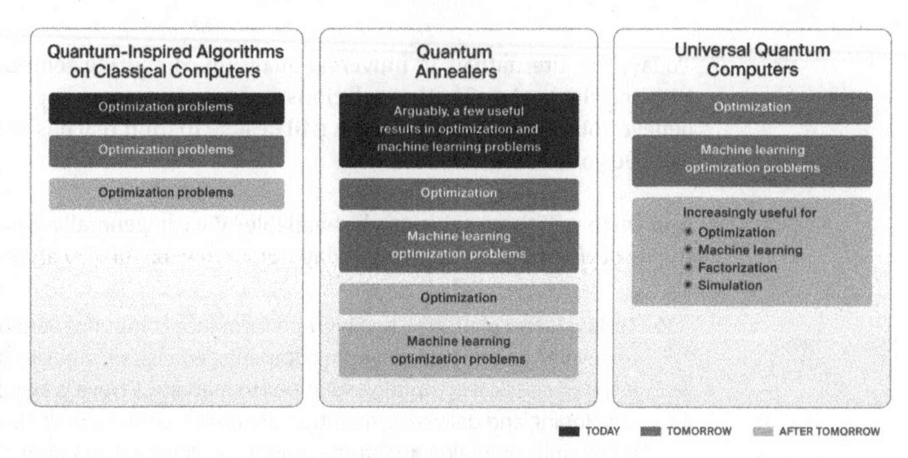

FIGURE 8-1: Expected usefulness of quantum computing solutions over time.

Ironically, a sharp slowdown in progress, called *quantum winter,* is something the industry fears and works hard to avoid. But AI, which is experiencing a huge spike in interest as we are writing this, went through several AI winters in the past, and may yet again. Even the internet, ubiquitous today, suffered from the dot com crash of 2001 and is currently again in a slump of venture capital activity and company valuations. So booms and busts in interest and activity are often part of the story, even for very successful technologies.

We will also add that, after "after tomorrow," this superposition of quantum computing approaches will probably collapse. (We're being literary here, not literal; no multiverse forking is expected to occur as a result of this particular collapse.) Universal quantum computers may become so powerful that there will be less need for quantum-inspired algorithms on classical computers, nor for quantum annealers and other analog quantum computers.

Dancing to That Algorithm

To fill the gap between most people's daily thinking and the algorithms that are so important to quantum computing, we have introduced a few of the leading algorithms, such as Shor's algorithm for factorization and Grover's algorithm for search, in earlier chapters (see Chapter 5 for brief descriptions of both these algorithms). Figure 8-2 shows an estimate of the speedup delivered when factoring large numbers by Shor's algorithm running on future quantum computers compared to classical algorithms (running, by definition, on classical computers).

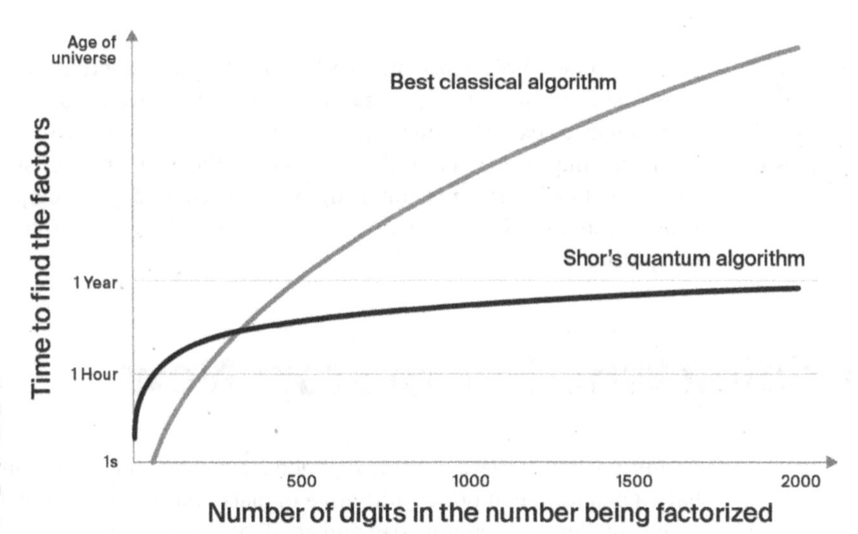

FIGURE 8-2: Factorization on classical computers versus a future quantum computer.

TECHNICAL STUFF

Shor's algorithm is in the upper tier in terms of its demands on quantum computing hardware. To run at the scale needed to crack current encryption standards, it will need many hundreds or several thousands of highly entangled, error-corrected qubits — systems that may not be available until after 2030. The reason that this seemingly remote prospect causes alarm in security circles is that cryptography needs are usually planned one to two decades out, and Shor's algorithm may come into usefulness in that time frame. Your humble authors believe that solutions will be ready in time, but it may be a close-run thing.

In this chapter, and elsewhere in the book, we talk about use cases, which is the practical work that quantum computing algorithms help you accomplish. In Chapter 14 we take you through a small bestiary of some of the most-used algorithms, to help you identify and apply algorithms to solve the problems you face in your work and to find algorithms that help you meet new opportunities.

TECHNICAL STUFF

If you're not mathematically oriented, you'll need a technical buddy or team as you put quantum computing to use. Such help can come from colleagues, consultants, platform vendors, and educational resources such as this book, and from software, system, and hardware vendors. Keep track of your questions as you go along so you can be sure to get them answered. And if you're already mathematically savvy, use this book and other resources to learn the language of business and various use cases, so you can help others put quantum computing to productive use.

Not everyone who's mathematically skilled is an expert programmer, and not everyone who's an expert programmer is highly skilled mathematically. Data scientists may be expert in math, programming, neither, or both. "You're technical" is an oversimplification that doesn't address the wide range of skills needed to get the most out of quantum computing. You'll need to hire, develop, or contract with people who are skilled in algorithm development to get the most out of quantum computing.

Deciding Whether to Start Now

At this writing, surveys indicate that about a quarter of large organizations already have a quantum computing initiative in place. Our impression is that the share is much smaller for medium-size and small organizations.

If you don't have an initiative in place, should you start now or wait? For large organizations, and for smaller organizations that depend on being up to speed on the latest technology, we recommend getting started now.

Being a little late to a party is one thing; not showing up until everyone else is heading home is quite another. Most organizations that are technically adept need to at least get their feet wet now, and this book is here to show you how to do that. As you learn more, you'll form a good idea as to how much further to go, and how fast.

On the other hand, it is indeed early days for quantum computing to be delivering significant value. Many organizations take this as a reason to wait. The idea is to hold off until quantum computing starts delivering solid value, and then get involved.

Machiavelli tells us that fear is a stronger motivator than desire. And this is often true with quantum computing. If you tell your management you can maximize returns with quantum computing, they might fund your project. But if you tell them that a competitor might get there first and put you out of business — the same kind of result but described in a different way — we believe, with Machiavelli, that your odds of success will go up.

WARNING

The threat of *Q day*, the day quantum technology makes today's internet security solutions obsolete, is a real concern. The video shown in Figure 8-3 is an example of the (sometimes alarmist) information available about this challenge. (You can find the video, Cybersecurity for the quantum era, at www.youtube.com/watch?v=udqoZHJJkBw.) Even if you don't believe you're ready to engage with quantum computing for competitive advantage, learn enough to be involved in discussions about quantum-based internet security solutions because many of these technologies are maturing rapidly.

FIGURE 8-3: The challenge to internet security posed by quantum technology is real, even if it's sometimes discussed in breathless tones.

In addition to gaining competitive advantage, some additional benefits to getting involved in quantum computing now are as follows:

- **Getting algorithmic:** Getting involved now helps you inventory and begin to improve the algorithms and applications that power your current business and new initiatives, so you can improve them across the range of quantum computing platforms as they develop.
- **Reaching a high ceiling:** If you can find a relatively small improvement due to your quantum computing work ahead of the competition, it can deliver a big payback.
- **Cashing in on compound interest:** Early traction may help you build a sustainable advantage, as early gains attract management attention, internal investment, and talent.

- **Getting prepared:** Early work in quantum computing helps you develop a strategy, develop and attract talent, and build teams and workflows, which are all needed to put quantum computing to work. If you start late, it will take a long time to catch up, if ever.
- **Developing relationships:** If you initiate relationships with providers and researchers now, you can grow them going forward. Whereas if you wait to move until it becomes commonplace, the people you need to work with may already be tied up.
- **Making a contribution:** In these early days, showing up with real business challenges and pushing for solutions helps the industry raise its game and make real progress faster. You become part of the development of the field and build strong connections.
- **Enhancing your employer brand:** Current and future employees love it when a company does cool stuff, and they love it more when good business results follow. Quantum computing is attractive to the very people cutting-edge businesses need most.
- **Rewarding your investors:** Investors, whether they be venture capitalists or owners of your company's stock, like to know that you're on top of what's going on in technology. And if you do gain real competitive advantage through quantum computing, all the better.

This book provides many ways to get started. Take the time to begin now so you don't get left behind.

Getting Your Organization Involved

"If you want to go fast, go alone; but if you want to go far, go together." — A proverb attributed to African cultures.

Organizations can get involved in new technologies, including quantum computing, in many ways. Here's how organizations of different sizes might find their way in:

- **Very large:** The largest organizations tend to have innovation centers or teams, under that name or some other name. Most such groups are already engaged with any technology that makes a top five or similar list of potentially game-changing technologies, and quantum technology — quantum computing, quantum sensing, and quantum metrology (measurement) — definitely makes it on any sensible list.

If you're not using or piloting quantum technology in general and quantum computing in particular, you're behind. Start by finding external quantum experts to help you identify some quick wins relevant to your business. Then choose key goals and identify what internal talent and external resources can help you meet them.

- **Large:** Large organizations that don't have established innovation centers or teams will often launch project teams to study new technologies and decide whether the organization should get involved.

  If you haven't already, it's time to formally assess whether quantum technology in general and quantum computing in particular are useful for your business. We describe how to put together a project team for this purpose in the next section.

- **Medium and small:** Organizations that fall in the small to medium enterprise (SME) category are vaguely defined but might include any business with more than ten or so employees or more than $1M in revenue. These organizations are selective in their embrace of new technology.

  You have to decide, on little information, whether it's worth investigating quantum technology for your business. Read this book, and then trust your gut.

- **Tech startups:** Tech startups can be anything from a single person with an idea to a unicorn (startup valued at more than a billion dollars). They are hungry for new ideas but have to choose their areas of focus carefully.

  You may choose to ignore quantum technology altogether — or to pivot your business to be all about it. Have the needed discussions now.

- **Small business:** Non-tech startups and established smaller businesses tend to rely on vendors and consultants for new technology.

  Learn enough about quantum technology to be able to have intelligent discussions with vendors, consultants, and current and potential customers about it. This book may be all you need.

WARNING

If you decide not to get involved right away, don't put quantum computing on the shelf for the next few years. Keep reevaluating, preferably annually, as the technology matures. Watch out for two things: opportunities you might miss, which is especially relevant for larger companies, and "category killers" that might put you out of business, which has extra relevance for small business. Think of the fate of many small bookshops as Amazon.com's online bookstore sucked up book sales in town after town.

ROADKILL ON THE INFORMATION SUPERHIGHWAY

One-time US vice-president and two-time presidential candidate Al Gore used to refer to the internet as the "information superhighway." And following the old rubric about innovation of any kind — that you can lead, follow, or get out of the way — those who failed to at least get out of the way were referred to as "roadkill on the information superhighway." The figure shows what can happen if you miss a technology 18-wheeler that's bearing down on you and your organization. Don't let yourself be roadkill on the quantum technology superhighway.

Along with evaluating whether a given new technology can help your business, you should also consider whether the technology might create a "black swan" threat, what risk managers refer to as a high impact, low probability (HILP) event.

For example, Barnes & Noble didn't realize that the internet, in the form of the Amazon.com bookstore, was going to decimate them. Freelance blog post writers and graphic designers didn't realize that ChatGPT, DALL-E, and similar tools were going to threaten their business. You don't want to be the one who didn't realize that quantum technologies, wielded effectively by others, might wipe out your business.

Take a look at your own business, and let your mind roam. Quantum sensing, for instance, may lead to better fish finders, which could improve your chances of catching a few fish on the weekend — and let big fishing fleets put some independents out of business.

Even big opportunities can become big threats, if competitors take advantage of them before you do. Most of the first movers on quantum technology are the largest businesses and government. If you're in a smaller business, you may have serious challenges from larger competitors — and equally serious opportunities from selling those technologies to other businesses and government.

If you periodically take a broad look at quantum technology and other fast-changing technologies, you'll be better able to surf the waves of change.

Putting together a project team

Larger organizations have to carefully assess new technologies before putting significant resources into them. Here we describe how a large organization can explore quantum computing in a way that lays the groundwork for serious engagement with the technology. If you're in a smaller organization, consider how to put together a scaled-down effort that accomplishes the same goals.

It's possible that you're the only person in your company interested enough in quantum computing to read this book — and if so, congratulations, because you're setting yourself up for success. But if you're in a medium-sized or large organization and there's interest, try to put together a small team to spend some of their time (not full time) looking into quantum computing.

On *Star Trek*, the Away Team is the group that beams down to the surface of a new planet to breathe the air, meet the inhabitants, and catalog the wildlife. (Or, if you're Captain Kirk, in many episodes, experience some wild life.) Not a bad metaphor for starting with quantum computing. (And the transporter would make a really good entanglement demonstration.)

So if you're starting a quantum computing effort, what are the roles that you would ideally have represented on the team? It's not that different from those on *Star Trek*:

- **Executive sponsor:** A C-level executive or one step lower.
- **Technical director:** A senior manager from a relevant technical discipline.

» **Algorithms expert (classical):** A techie who understands the IT team's work in terms of (classical computing) algorithms; might be a senior data analyst or an architect.

» **Algorithms expert (quantum):** A techie who knows quantum algorithms. You probably don't already have this person but you can look for them while you're exploring and vetting vendors.

» **Technical contributor(s):** One or more techies who may program or otherwise contribute to a proof of concept.

» **Business director:** A senior manager from the business side.

» **Business analyst type(s):** One or more business types who can help with the business justification around a proof of concept.

Note that this is roughly seven people. From our own experience, you don't want the team much larger than this, or it becomes too hard to organize meetings, reviews, and so on, and the whole effort will bog down.

If you're in a smaller company or just can't scare up much interest in a larger company, you may not be able to get all these roles involved. Be aware of who's not at the table and try to find the missing perspective wherever you can. The consulting company McKinsey notes that as few as three people may be enough to lead an initial effort.

And if it's just you, try to find a buddy to help. If you're technical, find someone with a business perspective and vice versa. If you can't get a pal, again, be aware that you don't have a broad view and try to fill in the missing viewpoint from research or other sources.

Figure 8-4 shows how technical and business advantages relate to your organization's overall goals. Use it to quickly classify opportunities that arise.

TIP

If you have to add more people, make them optional for meetings and reviews so that the increased scheduling difficulties won't slow you down. "Fire" them if they don't show up and contribute.

WARNING

If you have serious discussions about quantum computing with a group that doesn't include most or all these perspectives, even if they are not represented internally, the result is likely to be unbalanced from the point of view of the business as a whole.

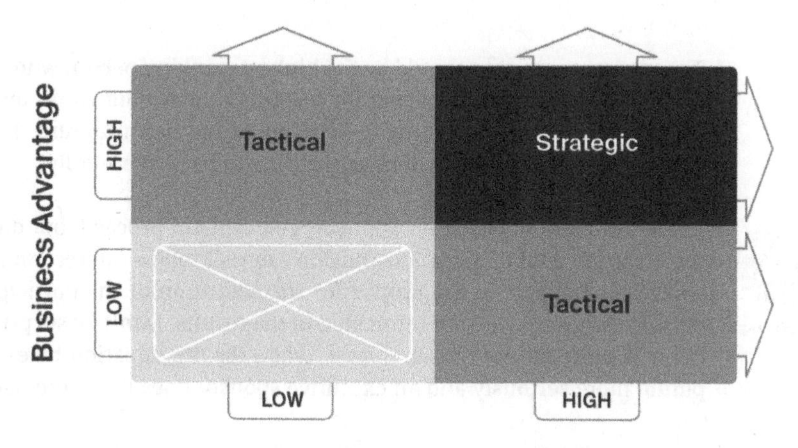

FIGURE 8-4: Identifying tactical and strategic opportunities for your organization based on the technical and business advantage you hope to achieve.

Involving an executive sponsor

Let's start at the top — of the org chart, that is. The role of the executive sponsor on any project team is critical. One might hope that good work by a project team that lacked an executive sponsor would result in attracting one later, but it rarely seems to work out that way.

An executive sponsor keeps the team aligned with the goals of the business and gives all of you "air cover" for the time, effort, and money you're spending. This person is critical to the effective operation of the team and even more so when it comes to the end of the initial project. An executive sponsor

- Helps shape the vision and how it fits into the company's overall roadmap and goals
- Gets attention for your findings within your company
- Obtains approval for any follow-up actions the team recommends, such as starting a proof of concept project

The executive sponsor doesn't have to lead the project but does have to be actively involved for the project to be successful. Ideally, the executive sponsor will see a lot of benefit to themselves for participating.

The executive sponsor should be the highest-ranking person who's interested and will make themselves available for meetings, reviewing work, and keeping their peers at the top of the company up to speed on your progress. The best "get" is usually the CEO, if they will make the time to participate fully.

WARNING

If you can't get an executive sponsor, you can still proceed, but don't expect your results to get a lot of attention and don't expect follow-up recommendations to be approved. Instead, do the project for the education of the participants and those outside the group who are interested in the results. Don't be surprised if the entire effort is more or less repeated later, when the organization takes quantum computing more seriously and an executive sponsor does make themselves available.

Depending on algorithm experts

Quantum computing depends on algorithms, so perhaps even more important than the executive sponsor are the roles of the two "algorithm whisperers" — one from classical computing and one from quantum computing. For the classical computing algorithm whisperer, you need someone from your own company who knows which classical computing algorithms are important to your company, understands your tech stack(s), and more. This person is vital for translating technical information to technical advantage. This person will then help the team as a whole determine where business advantage is possible.

In a perfect world, you will also have a quantum computing algorithm whisperer: a person in your company who understands existing quantum computing algorithms and can help you map them to your business needs. This person would be in active conversation with the classical-computing-based algorithms expert. However, you will probably need to buy or borrow this expertise, because few companies have it available internally.

WARNING

When you do hire outside experts, designate people in your company to learn from them, so significant knowledge transfer takes place. Otherwise, neither initial nor later efforts will pay off to the degree needed.

Here are some of the most widely used ways to tap into quantum computing algorithm expertise, and our rating of them:

- **Internal expert:** The very best. This person gets more and more useful as their knowledge and experience grow. If you have one such person, make it a priority to have them help you develop one or two more.

» **External consultant:** Very good. This person owes their loyalty to you. Ideally they already know quantum algorithms well and can learn your organization's needs quickly.

» **Cloud platform representative:** Good to very good. Plus points if their platform offers many different vendors. Be wary if their cloud platform has internally developed offerings to sell because they'll have trouble not being biased.

» **Vendor representative:** Good to very good. However, it's up to you to filter any bias the vendor may bring to the discussion. And when you're ready to invest more heavily, vendor reps will be crucial to success.

In early technology markets, vendor representatives can become a surprisingly important part of your network. Figure 8-5 shows a suggested path for engaging with them.

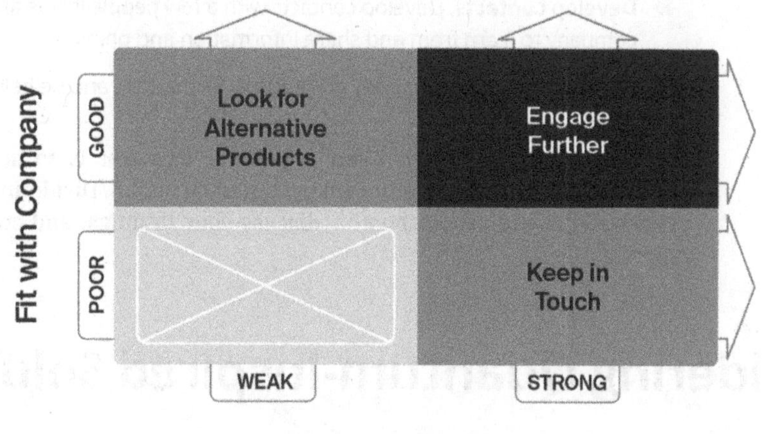

FIGURE 8-5: Future actions to consider after evaluating a product presentation.

TIP

If someone technically savvy wants to join the team and work toward becoming a quantum-computing-savvy algorithms expert, by all means encourage them. For example, some people might have studied quantum-related topics in school but ended up doing something else. And those with a physics background are likely to be good candidates for upskilling.

Setting goals

Many professionals work with a "crawl, walk, run" approach to taking on new technology. We're going to share some goals in the "crawl" category:

- **Learn the basics.** This book should help you develop the vocabulary and conceptual framework to make good use of further resources.
- **Be a point person for relevant news.** Use a Google Alert or similar to steer your way to stories relevant to your own expertise. Then develop a viewpoint about them, and share news and opinions with others.
- **Learn to evaluate vendor presentations.** Assess quantum computing vendor presentations from your respective perspective (such as leadership, business, or technical), ask appropriate questions, and make recommendations. Better still if you can attend a presentation alone and recommend whether something is worth the team's attention.
- **Develop contacts.** Develop contacts with a few people inside and outside the company to learn from and share information and opinions.
- **Show up.** Do reviews quickly on request, be positive, and be helpful.

How do you know if your team is gelling? One test is to attend a quantum computing trade show together (in person or virtually). Divide and conquer across presentations and vendor booths, discuss your findings, and create a brief joint report.

Considering Quantum-Inspired Solutions

In Chapter 1 we mentioned that part of the value of quantum computing to IT teams and the entire tech industry is that quantum computing inspires new approaches to use with classical computing systems. This benefit comes to life when you start looking into quantum computing for your company.

Ironically, many of the solutions that arguably demonstrate business value today are quantum-inspired algorithms running on classical mainstream or special-purpose hardware:

» **CPUs or GPUs:** An increasing number of quantum-inspired algorithms run on the central processing units (CPUs) found in every classical computer or the graphical processing units (GPUs) available for any classical computer.

» **Special-purpose hardware:** Hardware is available that's designed to run quantum computing algorithms to solve problems that feature the combinatorial explosions that typically stymie classical computers (see Chapter 1). Usually, this is supercomputer-type hardware; some new supercomputing systems have fast interconnects so that they can share problems with quantum computers in the future.

Quantum-inspired algorithms running on classical hardware are more likely to deliver value today than a true quantum computer, whether that's compared to a quantum annealer (Chapter 9) or a logic-gate quantum computer (Chapters 10 and 11). Neither type of true quantum computer may be mature enough, at this writing, to deliver value for your use cases today.

Using CPUs and GPUs

Quantum-inspired solutions running on CPUs and GPUs can be done on a do-it-yourself basis. Many organizations have their own data center, and any organization can get time in a public cloud. All public cloud vendors offer GPU-heavy configurations as needed.

Working with a vendor can be helpful because they can provide expertise on the specific problems their offerings are good at solving. If you want to work with a vendor, the chipmaker Nvidia, AWS through their Amazon Braket quantum computing initiative, and Azure Quantum all offer help with quantum-inspired algorithms running on CPUs or GPUs or both. (In each case, you're likely to be running the solutions you create on their hardware, so it makes sense for them to help you get started and be successful.) Strangeworks is also a good resource here.

Using special-purpose hardware

Special-purpose hardware is available from major high-performance computing (HPC) vendors, including Fujitsu, Hitachi, NEC, NTT, and Toshiba. Discuss your needs with several vendors, and then choose a vendor and contract with them to get started.

Special-purpose hardware is very powerful and can be very expensive. Expect vendors to have a good handle on how likely it is that their solution will provide value to you, but don't expect them to be heartbroken if you spend more than you initially anticipated.

Getting started with quantum computing in all its forms can be quite cheap, with emulators/simulators that run on your laptop and free credits for the use of various machines. But when you move toward serious workloads, quantum computing can suddenly get expensive. In the longer term, quantum computing will solve the problems that it's well suited to at a lower cost than classical computing and with less energy expenditure as well. For now, manage your efforts carefully to avoid sudden spikes in cost. Quantum-inspired solutions can be a good way to get real work done at less cost than today's fully quantum efforts.

Including quantum-inspired solutions

For a range of optimization problems running at scale, quantum-inspired solutions deliver business value today. These problems include those that can be stated as quadratic unconstrained binary optimizations (QUBOs), as explained in a general way in Chapters 9 and 14. If you have a use case that matches what these vendors advertise and have allocated heavy spending (several tens of thousands of dollars a year, to start), these vendors can help.

Quantum-inspired solutions are also an intermediate step on the way to fully quantum solutions. The mathematical representations you create for a quantum-inspired solution may work with little change on fully quantum systems as they mature. To use a baseball analogy, because up-front costs are lower and the solution complexity is less, you don't need to hit a home run; a ground-rule double will do very well indeed.

Use cases for these problems in the areas of chemistry and drugs include improving protein design, improving peptide design, searching for similarity between molecules, and optimizing clinical trials. Other use cases are supply chain optimization, route scheduling, shift scheduling, inventory management, and optimizing complex manufacturing cycles.

It's expected to be at least several years before logic-gate quantum computers deliver solid value. This time lag is largely due to the need to improve error rates while increasing qubit counts, along with other improvements.

When key thresholds are reached, logic-gate quantum computers are expected to do the following:

- Solve many optimization problems more effectively, at greater scale (eventually far greater scale) than quantum-inspired and quantum annealing solutions
- Solve many machine learning problems even better than quantum annealers
- Address problems in quantum simulation, including in chemistry and materials science, and also in cryptography, far better than any other kind of computing system

» Looking into annealing as an option

» Getting the right kind of quantum computer

Chapter 9

It's All about the Stack

Once you look at the opportunities in quantum computing available to you and your organization, you can start taking advantage of what's on offer. For business advantage today, you're most likely to find answers in quantum-inspired algorithms running on classical computers, as described in Chapter 8, and in using quantum annealers, a kind of quantum technology optimized for, well, optimization problems and showing promise in machine learning.

In this chapter, we introduce the stack, which powers all quantum computing solutions, and quantum annealers. In Chapters 10 and 11, we discuss the rich but confusing range of universal quantum computing systems. After we've introduced you to the range of options, you'll be ready to get the most out of quantum computing, today and going forward.

Analyzing the Stack

Like a classical computing system (see Chapter 2), a quantum computer is made up of a stack of technology — several layers of hardware, firmware (software semipermanently burned into chips), and software. The stack allows operations at one level of the system to be buffered from changes in another.

REMEMBER

Part of the reason for the effectiveness of quantum-inspired algorithms running on classical computers, as described in Chapter 8, is that the stack for those computers is optimized and mature. Moving to quantum computing hardware means leaving all that experience behind.

Examining how classical computers stack up

The layers of hardware and software in the stack make the entire system more reliable and more flexible. For instance, if you're running Microsoft Windows, you can move your applications and data to a new PC-compatible computer — widely referred to as a *Windows machine* — without much worry as to whether the applications and data you previously used will continue to work. The same is true when migrating within the Mac computer family that runs macOS.

In both cases, the operating system — Windows or macOS — supports a consistent user interface and thousands and thousands of applications, isolating them from most changes in the underlying hardware. Users need updates to their applications when major changes occur, but the applications continue to work in the same way and data migration is almost seamless.

The Mac family has gone through big changes in the 30 years that it's been around. In 1984, Apple launched the Mac using the Motorola 68000 family of microprocessors. They later moved to the PowerPC microprocessor series, then to the Intel x86 family, and recently to custom Apple silicon, in the form of the M1 and M2 microprocessors. (Halfway through this process, Apple also migrated Mac users from the original macOS, which Apple created, to a new OS based on the Unix open-source operating system, acquired from Next Computer when Apple bought that company in 1996 and brought back Steve Jobs.)

TECHNICAL STUFF

For historical reasons, Apple's Macintosh computers, which run macOS, are referred to by their family name as Macs. But IBM PC-compatible computers, which run Microsoft's Windows operating systems, are called Windows computers. To add complexity, both types of computers are often used to run the open-source Linux operating system, on its own or alongside the usual operating system. Linux is also widely used on cloud computing servers, supercomputers, and other systems. Recognizing the complexity in today's laptops and desktop computers may help you abide the much greater complexity that we see among quantum computing options today.

The hardware and software stack on a quantum computer works in much the same way — isolating the applications running quantum algorithms, for example, from changes in the underlying hardware. But the variety of quantum computing systems available now, and the changes likely to happen in the future, are far greater than anything faced by the Mac users we just mentioned.

TIP

Even when using classical computers and devices that include classical computing capability within themselves, the lower levels of the stack can make themselves known to you. For instance, for a smart device such as a digital camera, you may have been told: "You need a firmware upgrade." But even then, you just follow some steps. The functionality and capability of the firmware layer in your device are probably not significantly evolving over time, whereas for quantum computers, they are.

Comparing the stack for quantum computers

Here is the kind of migration that you, as an early adopter of quantum technology, may well undergo in the years ahead:

- **Quantum computing emulator/simulator:** You can start creating new software or modifying existing software on a quantum emulator/simulator running on your laptop or on classical computing servers via a public cloud provider. (As we mention in Chapter 8, quantum emulators — a software layer that allows software written for one kind of hardware to run on another — are often called *simulators* in the quantum computing world.)
- **Quantum-inspired software:** The most widely used way to get business value from quantum computing today is, ironically, to run quantum-inspired algorithms on classical computing systems, as described in Chapter 8. These solutions may offer varying degrees of improvement over non-quantum-inspired software running on classical computers, depending on your use case and algorithms.
- **Quantum annealers et al:** Quantum annealers are not universal computers and are limited to specific types of optimization problems, though a wide range of tasks can be handled within these limitations. Today's quantum annealers may soon offer incremental but often valuable advantages over quantum-inspired software. There's also a newly emerging class of other analog quantum computing systems that we don't cover much in this book.
- **Universal quantum computers:** Universal quantum computers today do not yet offer advantages over other systems for real-world business problems because of various issues, described in Chapter 10. However, you may want to start running algorithms on them today to get ready for the time when they do offer real advantages — advantages which will eventually, in many cases, be far greater than those offered by alternatives.

Cloud platforms such as AWS, Azure, and Strangeworks offer a range of options that cover the preceding kinds of solutions. Each cloud platform may offer options that don't appear on the others. AWS and Azure each offer superior integration with the AWS-specific or Azure-specific services offered on their respective platforms; Strangeworks offers the widest range of quantum computing alternatives, including access to both AWS and Azure offerings, and specialist knowledge of quantum computing solutions.

Even though several providers offer both kinds of computing, there's a big different between the classical computing stack and the quantum computing stack:

- **Classical computing:** The levels of the stack are so well worked out that you, as a customer and user, don't usually even need to be aware of them. They just do their job.
- **Quantum computing:** Every level is evolving and each level affects the others. You're likely to see frequent updates across different levels of the stack until quantum computing settles down and becomes more predictable.

Figure 9-1 shows a generalized stack for a quantum computing system. The stack you use for a specific system is likely to be very similar, but your mileage may vary.

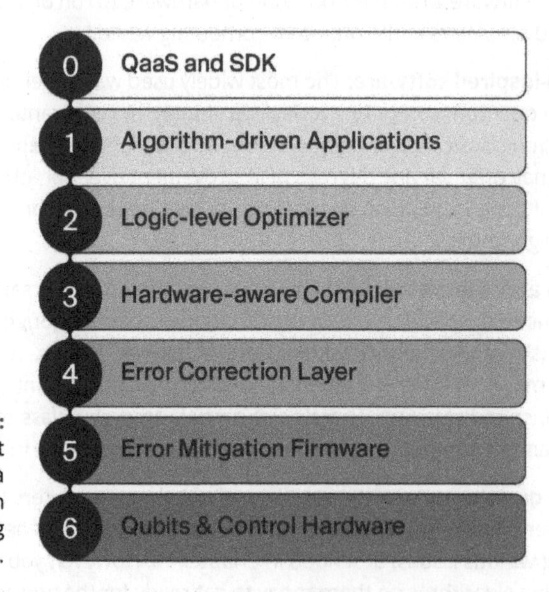

FIGURE 9-1: The layers that make up a quantum computing system.

The software layers are roughly the same across quantum annealers (with one qubit type and one manufacturer, D-Wave) and universal quantum computers (with five qubit types and more than a dozen manufacturers). But different vendors may use different names for the same layers or have stacks made up of different layers.

Qubits and control hardware

If you want to demonstrate to a classical computing expert that it's still early days in quantum computing, just tell them there are still more than a half dozen different kinds of qubits. They might say, "Oh, like the old days, when Intel made x86 microprocessors for IBM PC-compatibles and Motorola's 68000 series powered the early versions of the Mac?" And you could tell them, "Yes, but each qubit is a totally different technology. And they all access the multiverse, but each does so in a different way." (Whereas the PC and Mac microprocessors use the same underlying technologies, have the same kinds of surrounding hardware and firmware, and concern themselves with only one universe.)

But we're just talking about qubits here, so we'll restrict ourselves to saying that there are five kinds in commercial quantum computers today: superconducting qubits, trapped ions, photonics, neutral atoms, and quantum dots.

Each kind of qubit has different control hardware, including devices used to program the qubits and devices — sometimes the same device, sometimes an additional device of the same type, and sometimes a different kind of device — to read out the results. For instance, superconducting qubits use the same tiny lasers to program the qubits and to read out the results.

Error mitigation firmware

Qubits are sensitive to errors caused by interactions with their environment — temperature fluctuations, cosmic rays, magnetic fields, and vibration are the main culprits. Even without environmental interference, qubits tend to degrade toward decoherence fairly rapidly.

Firmware is software burned into chips that are part of the system. It can manage the environmental influences that qubits face so that the system works more stably and can run software with fewer errors and for a longer period. Firmware can work in partnership with an error correction layer to further improve software performance.

Mainframe computers and minicomputers used to be installed on raised flooring with shock absorption to guard against vibration that could, for instance, crash a hard disk drive, which could cause days of downtime and cost tens of thousands of dollars to fix. The term *computer bug* came about because a moth flew into the circuitry of an early mainframe! But the operational challenges facing quantum computers today still far exceed those that classical computers have, for the most part, overcome.

Error correction layer

Most quantum computers available today lack an error correction layer, which can combine hardware and software, but this level is one of the most important areas for progress in the years ahead. Many techniques are available for use, including statistical suppression of errors and hardware-based error control.

The gold standard is the use of logical qubits, in which the information that ideally would be contained in a single, accurate physical qubit is instead smeared across multiple error-prone physical qubits, which is what the industry can actually offer. A good fly-on-the-wall guesstimate is that it may take anywhere from ten physical qubits to scores of them to create a single, fully reliable logical qubit. Figure 9-2 is a conceptual diagram of a potential mapping process for using several physical qubits to make up a single logical qubit.

Universal quantum computers are still limited to a few score or a few hundred physical qubits, so combining ten or more precious physical qubits to yield a single logical qubit is a big sacrifice of potential processing power. But, at this writing, that's the plan.

Phase-Flip Error   Bit-Flip Error

Data qubit   Measure qubit   Data qubit with error   Unused

FIGURE 9-2: E pluribus unum (out of many, one) — good for a country but hard on your qubit count.

Hardware-aware quantum compiler

On both classical and quantum computers, software programs are written in a programming language and compiled into machine language — code that directly tells the hardware what to do. In classical computing, each microprocessor (which has its control hardware built in) has its own machine language, which is well established. In quantum computing, each combination of qubits with their associated control hardware has its own machine language, and it's subject to frequent revision as the technology changes.

The lower-level compiler is very important on quantum computers because the lower layers of the stack we're describing here are still changing quite a bit, and the underlying hardware is buggy and not error-corrected.

A hardware-aware quantum compiler can minimize the number of hardware steps needed to complete each programming language instruction, improving the odds that the program can finish before decoherence sets in and before errors occur.

TECHNICAL STUFF

Assembly language is a programming language that's one step up from machine language; as such, it's different for every microprocessor. One of the authors (Smith) got to program the 6,502 microprocessors that powered the Apple II in assembler, as it's called. This required working in agonizingly small steps to, for instance, display a horizontal line of text on the screen. It was a blast. If you're interested in pushing the limits of performance today, you may get more out of your quantum computing efforts if at least some of your programming work is at the lowest level available for the specific quantum computer you're using. (And yes, this may require a physicist or two to optimize the lowest level of the quantum computing value chain.)

TECHNICAL STUFF

Moving the operating system and applications from one Mac system to another doesn't require rethinking your basic assumptions about computing — or about reality. The CPUs on the two systems are likely to be compatible, with a newer CPU perhaps being faster or offering a few additional machine language instructions. One computer may have more memory than another or different peripherals. But on quantum computers, important assumptions can change. For instance, the low-level code that runs well on a quantum annealer may not work nearly as well on one type of universal quantum computer or another. You'll have to continually question whether each level of your stack is providing the device independence and software optimization that you would like it to as hardware evolves and as you work on different kinds of systems. And you'll need to benchmark different implementations of your software against each other to make sure you're getting the most out of each different kind of system.

Logic-level compiler and optimizer

A logic-level compiler takes the software code written by a developer and prepares it for the hardware-level compiler. A logic-level compiler is more or less device-independent; its output is the same (or close to the same) no matter what hardware it's running on.

A program for a gate-based quantum computer is called a *circuit*, which is a series of logic gates applied to a qubit to run an algorithm. We provide details and examples in Chapter 12.

To accomplish the same result, you can write a program in a lot of different ways. An optimizer is meant to take whatever the heck you've written in your code and reorganize it into a more efficient form.

The circuit can be run against logical qubits, which reduces errors, or physical qubits, which gives you more qubits to work with but increases the chance of an aborted program (er, circuit) or errors in your results.

Algorithm-driven applications

To a much greater degree than classical computer programs, quantum computing programs are straightforward implementations of selected algorithms. That's because classical computers are the more convenient choice for most tasks; you bring in the quantum computer only when you have a narrow task that can be completed (or partially completed) faster when mapped to specific quantum algorithms, or when you're experimenting and learning.

For information on programming a quantum computer, see Chapter 12; for information on algorithms to use in programming, see Chapters 13 and 14.

Cloud-based quantum service and software development kit

Most people don't have a direct connection to a quantum computer and instead use a cloud-based quantum service, whether through a public cloud provider (such as Amazon Braket), a quantum computer's cloud interface (such as IBM Quantum), or a quantum computing portal such as Strangeworks, the company run by whurley, co-author of this book. These are all called quantum (computing)-as-a-service (QaaS) offerings.

You're likely to be writing software in a software development kit (SDK) that works with an existing language such as Python, resembles an existing language such as C or C++, or is quantum computing specific. You can also go even lower-level and use Open Quantum Assembly Language (OpenQASM, pronounced "open chasm.")

In these early days, carefully consider the option of making a quantum-computing-specific language or OpenQASM part of your toolkit, or having someone with that skill on your team. If not, vendors will typically have expertise you can leverage. You'll want to optimize the sweet living heck out of any program you get actual economic value from, to maximize that value, and using a quantum-computing-specific language can be an important part of that process.

Considering the Annealing Alternative

Quantum annealing is a different type of quantum computing that is currently only on offer from one vendor, D-Wave Systems (the first quantum computing company founded, in 1999). A recent D-Wave system — the D-Wave Advantage, with 5,000 qubits — is shown in Figure 9-3.

Quantum annealing solves only one type of problem, called an optimization problem. However, many quantum computing problems fit this description, including the traveling salesperson problem, which we've mentioned several times. Here's another example of an optimization problem: Start with the stock price movements for companies in the Dow Jones Industrial Average of stock prices over the last 20 years. Create an algorithm to choose a 10-stock portfolio that, historically, has had the lowest risk and the highest reward. Unfortunately, past performance is not a guarantee of future results, or making money this way would be easy. But stock-picking algorithms using quantum computing are having some initial success today.

The qubits in annealers aren't quite as sophisticated as the qubits in universal quantum computers. We explain this topic in more detail in Chapters 10 and 11, which consider several kinds of qubits in some depth.

FIGURE 9-3: A D-Wave computer used to solve an optimization problem.

D-Wave Systems Inc.

REMEMBER

Even computers described as universal computers are not as universal as we would like them to be because the tape in the theoretical Turing machine is infinite and any read/write store available for use in this dispensation is finite. This difference causes a lot of interesting bugs in computer programs written by us finite carbon-based units when RAM or disk space runs out of capacity or when the network stops working. But we still use the term *universal computer* for devices that can perform any imaginable set of steps or mathematical calculation, with the caveat that they have the needed resources.

The annealing technology is worth reviewing carefully, even if you intend to use a universal quantum computer, for several reasons:

- Quantum annealing is more mature, with a several years' head start on universal quantum computing.
- Some types of problems may always work best on a quantum annealer, so it's good to have annealing in your mental toolkit.
- Quantum annealing is more limited than universal quantum computing and therefore easier to learn.

» The logic behind annealing is easier to explain than with other types of algorithms, while serving as a good example of what's involved in algorithms as a class.

» What you learn about computing problems when implementing annealing also applies, in large part, to classical computing and to universal quantum computing.

Quantum annealers solve a lot of interesting problems and can be a solid step into the broader world of quantum computing. Cloud access makes it easy to compare the available options and choose the right one for your use case.

TOUGH LOVE FOR D-WAVE'S FIRST SYSTEMS

When D-Wave introduced their systems publicly in the late 2000s and shipped commercial systems in the early 2010s, there was huge interest in quantum computing, as there is now. But D-Wave's systems are quantum annealers, not fully capable universal quantum computers, so the community reacted harshly to their introduction.

At first, there were questions as to whether D-Wave's systems were really computers because they don't meet the Turing machine standard described in Chapter 2 — that is, they can't solve any conceivable logical or mathematical problem. True enough, but a device can be a computer and not meet that standard, though classical computers and universal quantum computers do.

Then there was the question as to whether they were really quantum computers or just classical annealers dressed up in quantum clothing. D-Wave's machines have withstood the test here, and are generally agreed to be a type of quantum computer.

Then there were concerns about performance. Does a quantum annealer offer sustainable advantages over classical computing? Quantum-inspired algorithms running on classical computers (see Chapter 8) have raised some doubts here.

And finally, D-Wave's systems use a limited form of entanglement amongst qubits, so they don't scale to the same (incredible) degree as a fully capable universal quantum computer. D-Wave addressed this by throwing (somewhat lower-quality) qubits at the problem; their highest-end systems have roughly 5,000 qubits, whereas IBM's high-end universal quantum computing system has somewhat more than 100. (This is not anything like a 1:1 comparison, however, because quantum annealers can't solve all the same problems that universal quantum computers, such as IBM's, can.)

D-Wave has recently announced that they intend to build a universal quantum computer in the years ahead, so everyone will see what they bring to both arenas.

TECHNICAL STUFF

Quantum-inspired algorithms running on classical computers, as described in Chapter 8, solve roughly the same kinds of problems as quantum annealers. When it comes time to run your problem, you may find that one of these types of computation offers cost, performance, or reliability advantages over the other.

TECHNICAL STUFF

A quantum annealer is also called an *adiabatic quantum computer*, where *adiabatic* means "without gain or loss of heat." The premise of quantum annealers is that they attempt to compute by moving the qubits toward the lowest energy state. The constraint in place as the program proceeds is that qubits are not allowed to gain heat (that is, move to a higher-energy state). For quantum mechanical reasons (see Chapter 3), a quantum particle that is already at its lowest-energy state cannot lose more energy.

Annealing in the medieval mode

Annealing is originally a process in ironworking. Most iron has a somewhat random internal structure, which tends to make it both stiff and brittle. Annealing makes the structure more regular and stronger. Figure 9-4 shows the change that annealing creates.

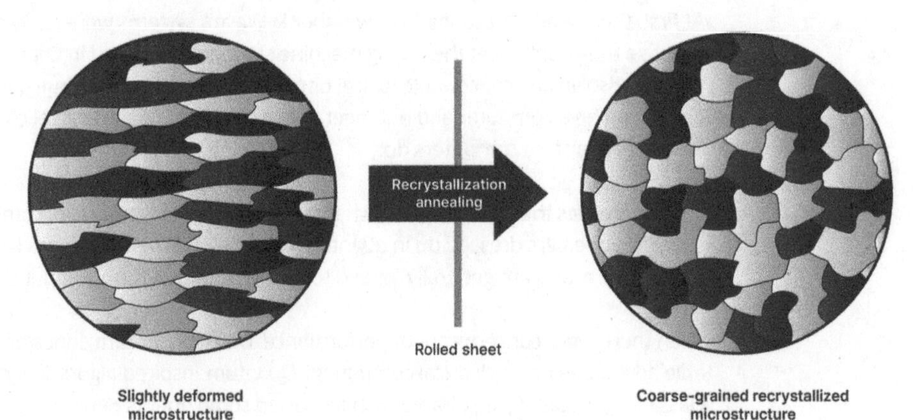

FIGURE 9-4: A representation of the structure of annealed iron.

If you heat iron until its internal structure dissolves and hold it at that higher temperature for some period of time, the internal structure breaks down. If the iron is then allowed to slowly cool, its internal structure crystallizes in a more orderly way, making for a softer, less brittle, more easily worked metal.

TECHNICAL STUFF

People have been doing annealing at least since the Middle Ages, without a deep understanding of the physics involved. As they said back in the day (the year 1400), "Take þe plates of bras pannes or of cawdrouns and anele hem in þe fire rede hoot" ("Take the plates of brass pans or of cauldrons and anneal them in the fire red hot"). Extra points for implementing that in Python.

WHY CAN'T A QUANTUM ANNEALER DO MORE?

The race in quantum computing often seems to be all about qubits. So why aren't D-Wave's quantum annealer systems, with more than 5,000 qubits, blowing the doors off universal quantum computers with somewhere around 100 qubits?

To sum up, the D-Wave qubits are of somewhat lower quality; they're limited in how much they can be entangled; and the underlying technology, quantum annealing, does not maximize the value of every qubit the way universal quantum computing can. In addition, no quantum computer so far has robust error correction, so all of them incur undetected errors and often deliver approximate, or simply wrong, results.

D-Wave, with quantum annealers, and providers of quantum-inspired algorithms on classical computing hardware (Chapter 8) are overcoming these negatives by providing many, many qubits.

It will take more time to determine the true strengths and weaknesses of each kind of quantum computing. What's already happening is a fascinating study in different approaches to the technology.

Identifying the problem annealing solves

Annealing metal brings it closer to its equilibrium state; it eliminates internal stress within the metal. Similarly, mathematical annealing brings a system to an equilibrium state — toward a minimum or maximum value.

Optimization problems are usually about finding the lowest or highest attainable value for a variable: the lowest cost, the smallest amount of material used, the shortest time to completion, the highest quality, the greatest profit, or the longest that a material can last under normal use. Many problems in our daily lives are optimization problems, and so are many problems in computing.

TECHNICAL STUFF

The key value in a quantum computing algorithm is often referred to as the *cost*, but this is not necessarily a cost in dollars; it may refer to maximizing or minimizing computation time, calendar time, physical effort expended by workers, or some other variable.

Optimization problems usually have constraints as well. The lowest possible value is often zero — nothing is cheaper than free; the shortest time you can get a desired result is no time at all. The highest allowed value may reflect real-world constraints: few people will pay more than $100,000 for a car; few people want to wait more than 10 minutes for their hamburger to be served to them.

The easiest way to find a minimum (or, turning things around, a maximum) is the hill-climbing method:

- Measure the desired variable at a point on a function.
- Measure the same variable at a nearby point on the function.
- Compare to see whether the value of the variable has increased (moved toward a maximum) or decreased (moved toward a minimum).
- If the new answer is better, move the same distance in the same direction and try again; if it's worse, move the same distance from the original point in the opposite direction and try again.
- Repeat until the highest (maximized) or lowest (minimized) value of the variable is found.
- Consider trying some smaller steps to see if you can get a more precise answer.

If you look for a maximum or minimum value for a function that only has one high or low point within your constraints, the hill-climbing function will find it. But what if you try hill-climbing along a stretch where there are two or more hills? Figure 9-5 illustrates this complexity. A hill-climbing algorithm will tend to find the local maximum if it starts at points A or B, or the global maximum if it starts at points D or E. If it starts at C, the odds of finding the local maximum versus the global maximum are 50/50.

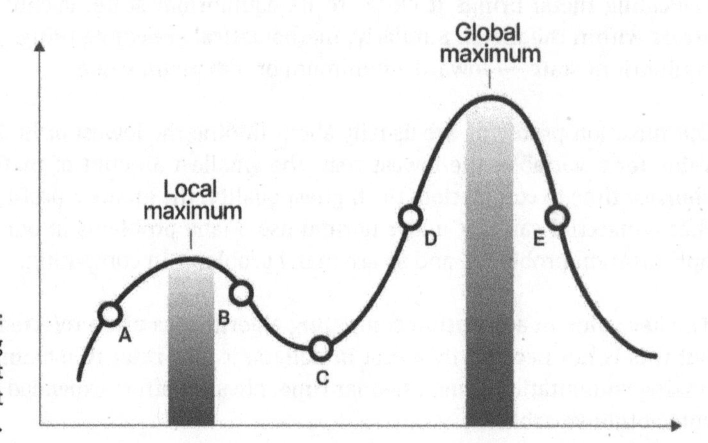

FIGURE 9-5: Finding the highest maximum is not always easy.

But there's a problem here. The hill-climbing algorithm may find you a hill while missing a mountain. This problem reminds us of the classic Hugh Grant comedy, *The Englishman Who Went Up a Hill but Came Down a Mountain* (1995) — but we digress. Again.

You can solve the problem of finding the global maximum by trying the hill-climbing algorithm at multiple points within your constraints, but this approach gets computationally expensive. And if you don't know how many hills there might be, you need to try a lot of times to raise your chances of finding the tallest one.

Let's say you live on a hilly street. You step out your front door and, with each step, you head uphill, whichever direction "up" is.

You will eventually arrive at some kind of hilltop or high point — a local maximum. But it won't be a global maximum unless you started on the slopes of Mount Everest, the highest point on Earth. Otherwise, you may attain a natural high, but you're likely to be many miles away from the global maximum on the border of Nepal and China.

For instance, in Figure 9-5, if you start at point A or B and walk uphill, you will end up at your local maximum. Only if you happen to live on the slopes of Mount Everest — point D or point E — will you end up at the global maximum, because it's also your local maximum. And if you start at point C, your destination depends on whether your first step uphill is to the left or to the right. (The classical algorithm is dumb in the sense that it has no particular attraction to the right, or the wrong, answer.)

So classical algorithms tend to get stuck with local, or at least non-global, maxima/minima. A quantum version can often overcome this problem.

Moving on to quantum annealing

Quantum annealing (QA) works differently than simulated annealing on a classical computer, in two particulars:

- Quantum annealing leverages the physical changes in the energy levels of qubits to find the optimal mathematical solution.
- Quantum annealing can tunnel to the desired global maximum/minimum, bypassing local maxima/minima.

Referring to Figure 9-5, if you start at point B, which is just downhill from a local maximum, QA will still ignore the local maximum and move toward the global maximum. And if you start at point A, QA will, if everything is working properly, tunnel through the hill represented by the local maximum and find the global maximum.

TECHNICAL STUFF

Quantum annealing is based on careful mathematical work by Tadashi Kadowaki and Hidetoshi Nishimori on optimization of annealing using quantum technology. While quantum annealing can't leverage the large mathematical knowledge base around universal computing in general and universal quantum computing in particular, it rests on solid theoretical work.

TECHNICAL STUFF

Simulated annealing doesn't take advantage of the physical properties of bits to do its work; it's simply a mathematical process running on a physical substrate provided by the classical computer. But quantum annealing does use the physical properties of qubits — in particular, the capability of quantum particles to tunnel around or through barriers by reappearing in random locations, moment to moment. (This movement is impossible in classical physics.)

Identifying where quantum annealers succeed

Quantum annealers are good at solving optimization problems, and not just annealing problems. A quantum annealer can solve any problem that can be stated as a quadratic unconstrained binary optimization (QUBO) expression.

The heck you say? We explain QUBOs in more detail in Chapter 14. For now, just focus on the word *optimization*; in commonsense terms, that's what quantum annealers are good for.

TECHNICAL STUFF

Quantum annealers can also solve problems presented using something called an Ising model, but it's easy for skilled practitioners to convert problems between an Ising model and a QUBO.

Also, know four things:

- Many problems can be stated as QUBOs.
- Many other problems can be reformulated as QUBOs, with some effort.
- Yet other programs can be more or less closely approximated by a QUBO, so you can get an imperfect but still useful answer, or at least a directional answer, from a QUBO.
- Some problems can't be restated as a QUBO, or at least not effectively. Among these problems are search (Grover's algorithm) and prime number factorization (Shor's algorithm).

For example, when D-Wave introduced their first system in 2011, part of the criticism centered on the fact that it couldn't execute Shor's algorithm, which is still the most consequential algorithm in quantum computing. There has been some comment that a QUBO-friendly solution has been found, but it's yet to be seen if it works at a level that results in advantage over classical computing.

Shor's algorithm uses the capabilities of quantum computers to quickly find the prime factors of very large numbers. Unfortunately for the cybersecurity community, the most widely used encryption algorithm, RSA, uses the difficulty of finding primes as its basis. So at some point, quantum computers are likely to be able to break most of the encryption in use today. See Chapter 11 for more information.

Quantum computing works in two ways. The first is by directly simulating the activity of quantum particles in applications such as materials science and drug discovery, which is what Feynman originally called for in 1959 (see Chapter 5). You generally can't do simulation with a quantum annealer. The second is via a kind of extended metaphor where logical or mathematical processes are mapped onto changes in the physical state of qubits, such that the end state of the qubits represents the answer to a problem. Annealing is a good example of the latter dynamic; a problem is stated such that the optimal answer maps to the lowest-energy end state for the qubits.

Identifying where quantum annealers fail

Many problems that a universal quantum computer can solve are not able to be solved, or nearly solved, as QUBOs. Chapter 14 describes which algorithms work for different kinds of problems.

Speaking broadly, quantum annealers don't give the system the same control per qubit that universal quantum computers have. They also have limitations on their capability to entangle large numbers of qubits. Therefore, they don't function well as simulators for quantum mechanical processes, making them not that useful for problems such as materials science and drug discovery — both problem areas that depend on simulations. There are also many kinds of problems that require logic gates of one kind or another. Logic gates of certain kinds are found in classical computers; a somewhat different set is used in quantum computers. But problems that require logic gates do not work on quantum annealers, nor on quantum-inspired classical computers (see Chapters 10 and 11).

Choosing a Type of Quantum Computer

One of the benefits of quantum computing is that it's making everyone look at what each type of computing does best — classical versus quantum and quantum annealing versus universal quantum computers — and how to get the most out of them all.

If you think that all you need to worry about right now is AI and machine learning, not quantum computing, that doesn't get you off the hook. Classical computing, including quantum-inspired algorithms on classical computers, quantum annealers, and universal quantum computers may all have important roles to play in maximizing AI.

With that in mind, let's summarize the relative disadvantages of quantum annealers and quantum-inspired algorithms on classical computers (Chapter 8), which solve similar kinds of problems:

- **Not universal:** Unlike classical computers and universal quantum computers, a quantum annealer is not a universal computer (see Chapter 2). It can run only optimization problems.
- **Poorer qubits:** Quantum annealer qubits are less complex, with simpler control circuitry, than those in universal quantum computers. They are also less powerful and have limited entanglement capability, but are easier to make in large numbers.
- **QUBO-only:** Quantum annealers run only problems that can be expressed as a quadratic unconstrained binary optimization (QUBO).
- **Not exact:** Quantum annealers tend to have lots of qubits but, like other quantum computers, little or no error correction. You're likely to get an approximate, rather than an exact, answer.
- **Hard to simulate:** Universal quantum computers can, by definition, be simulated mathematically, reducing the need for testing. With a quantum annealer, simulation is not always possible, and you just have to try stuff until you either do or don't get a useful answer.
- **One manufacturer:** At this writing, D-Wave is the only major manufacturer of quantum annealers, which reduces competition on price, features, and more. However, several providers are offering quantum-inspired algorithms on classical computers, which solve similar problems.
- **Less research:** There's a space race for the best universal quantum computer and the best kind of qubits. There's less research and activity around quantum annealers, and they use only one type of qubit (as we describe in the next chapter).

Some people whine because they have to learn about QUBOs or Ising models or both to use a quantum annealer. Sorry, Charlie, as the old commercials for canned tuna fish used to say: You, or someone on your team, needs to make good decisions about algorithms to get the most out of quantum computing of any kind.

So now for the good news:

- **More qubits:** Quantum annealers can use simpler qubits with less control circuitry; they're still magical, but in a slightly different (and easier to manufacture) way. So quantum annealers tend to have lots of qubit power.
- **More experience:** D-Wave got an early start, and providers of quantum-inspired algorithms on classical computers can draw on deep wells of knowledge, so there is a lot of experience with this kind of computing.
- **More focused:** The community around quantum annealers is dedicated to a specific set of problems, so they can share a lot of knowledge around doing those things well.
- **Flexible boundaries:** A surprising number of problems have been converted to QUBO form, and more are being added every day.
- **Important part of your toolkit:** Everyone with a wide range of computing needs will have some QUBOs in the mix, so they will likely need quantum annealers.
- **Good learning tool:** Quantum annealing is a good place to start. You'll learn the more constrained environment quickly and may be able to start getting results that are valuable for use in your business fast.

Theoretically, adding one qubit to a universal quantum computer doubles its power — but the lack of error correction and other problems with current quantum computers means that this improvement is still more a promise than a reality. Adding one qubit to a quantum annealer does not have the same effect, and it may turn out that a quantum annealer needs to double its qubit count to double its power.

We've summed up some key differences between annealers and universal quantum computers in Table 9-1. This is not a detailed spec sheet; things are changing too fast for us to provide that, and the two types of quantum computers are so different that specifics are a bit beside the point. The idea is to orient you as to what to expect from each type of system, now and in the longer term.

Please note that quantum-inspired algorithms on classical computers (see Chapter 8) and quantum annealers can be used for the same kinds of problems. Which is the better choice often comes down to the cost of the (classical or quantum) CPU time needed to generate a solution. The relative cost of using each kind of system to solve a specific problem is changing, so you'll probably need to research the best approach for solving your optimization problem when you're trying to solve it.

TABLE 9-1 **Comparing Quantum Annealers and Universal Quantum Computers**

	Quantum Computer Type	
	Annealer	Universal
Universal computer (in math terms)	No	Yes
Can model mathematically	No	Yes
Research activity	Limited	Robust
Functionality	Optimizer only	Any operation, including optimization
Algorithm type	QUBO/Ising model only	Many
Manufacturers	D-Wave	Many
First commercial access	2011	2016
Current number of qubits	5,000-plus	100-plus
Qubit types	One	Five
Entanglement?	Limited	Robust
Learning curve	Moderate	Hard
Public company	All (D-Wave only)	Some (IBM, IonQ)
Community support	Robust	Varies

» Winning with qubits

» Taking a stroll through the qubit zoo

» Finding reality in modalities

» Choosing a way forward

Chapter 10

Racing for the Perfect Qubit

The race for the perfect qubit is the Space Race of our time.

The peak years of the Space Race began in 1957 with the launch of *Sputnik*, the first satellite to orbit the Earth, by the Soviet Union, backed by the Eastern Bloc and China. The Soviets followed with the first animal (a dog, Laika), the first man (Yuri Gagarin), and the first woman (Valentina Tereshkova) in space. And the Soviets conducted the first-ever spacewalk in 1965.

The launch of *Sputnik* was terrifying to Americans, the citizens of Western Europe, and the rest of what was then called the free world. This fear arose because the Space Race was an outgrowth of the worldwide arms race, which saw the US and USSR rapidly building up their nuclear weapon capabilities. There was much talk of throw weights and kill ratios. The single most frightening moment in this all-out arms race, the Cuban Missile Crisis of 1962, came close to causing nuclear war.

The previous year, President Kennedy had promised that the US would "put a man on the moon in this decade." A global competition followed, pitting the military-industrial complex of each side against the other. A series of missions, including several disasters, led to the climax of the Space Race: the launch of *Apollo* 11 in July 1969, which put Neil Armstrong and Buzz Aldrin on the surface of the moon and cemented US leadership in space.

If you're interested to learn more, you can choose from many books, movies, and TV series. *Space Race* by Deborah Cadbury (Harper, 2007) is a good narrative history (see Figure 10-1). The Apple TV+ series, *For All Mankind*, is a fictional work of alternate history in which the USSR gets to the moon first.

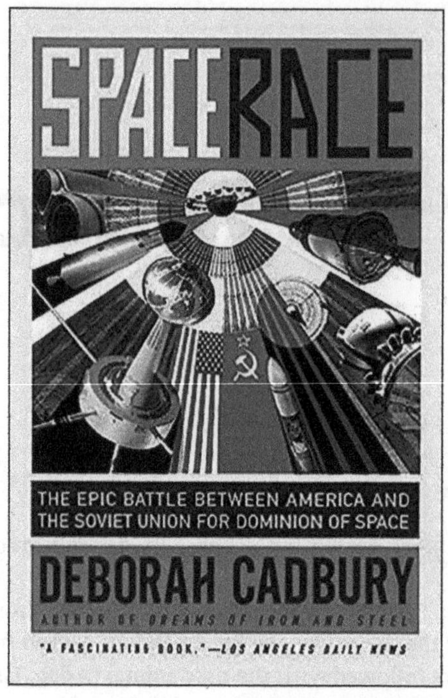

FIGURE 10-1: *Space Race* is one of many histories of the US versus USSR race to space and to the Moon.

HARPERCOLLINS PUBLISHERS

The Space Race led to huge advances in military capabilities — which, luckily, were never tested in all-out war — and in the technology available initially to the West and eventually to the world. Computational capabilities advanced rapidly, and in the US, ARPAnet — the precursor to today's internet — appeared in 1969. The satellite networks that power global telecommunications and the GPS system come out of this period, and the entire Quantum Technology 1.0 roster (see Chapter 5) — solar power, microwave ovens, medical scanners — was either invented, or rapidly improved, as part of the Space Race.

Today's immense race — both public and private — is in developing usable qubits and taking leadership in quantum computing and quantum communications. This competition spans companies and countries. Like the Space Race, the race for quantum computing leadership has national security and military implications, due to the near-certainty that Shor's algorithm (see Chapter 6) will make the military, governmental, and business secrets of today vulnerable to the quantum computers of tomorrow.

A fierce competition exists between the US and China in quantum computing, though scientists from all countries also continue a strong degree of collaboration and sharing. China has announced $10 billion in funding for quantum computing, but this represents most of China's total commitment, commercial as well as governmental and military. The US has announced more than $1 billion in governmental funding, is spending unknown other monies in secret, and benefits from billions of dollars of announced and unannounced investment from governments, companies, and private investors in North America, Europe, and around the world.

Progress in quantum computing is also tremendously important to businesses and individuals. To understand why, just reflect on the effect of the rise of the internet and the web since the early 1990s, as well as the large and fast-growing effect that AI and machine learning (ML) are having today. (One of the most important effects of quantum computing will be its mutually beneficial speedup of and interaction with AI and ML.)

Quantum computing will have a strong effect first on specific industries and then on society as a whole. And advances in AI, ML, and quantum computing are likely to accelerate each other, with results that no one can predict at this point.

THE WONDER OF QUBITS

As we dive into the practical and historic implications of progress in qubits, we urge you to keep in mind how amazing qubits are. We're using wild bits of living physics that we don't even fully understand to power computer algorithms that, for the most part, haven't been invented yet.

This task is enormously difficult, and rightly so. Keep this difficulty in mind. It will give you patience for the challenges of using the technology as it develops and for the many years that it will take to reach anything resembling its full capability. The difficulty of the task is directly related to the potential that is likely to be unleashed as various milestones are achieved.

As you engage with quantum computing, use your imagination and creativity to find new ways to put this steadily increasing power to use. As with the internet, and again with AI and ML, some of the ideas that sound craziest at first may be the ones that result in the best and most powerful innovations as quantum computing moves ahead.

So the race for the perfect qubit is a big deal. Many barriers must be overcome, including the need to move from a couple of hundred qubits today to many thousands; short qubit coherence times that limit the size of software programs; and the lack of error correction. In this chapter and the next, we describe why qubits are so amazing and what is needed to make useful qubits and combine them into increasingly powerful quantum computers.

Identifying Three Levels of Qubit Achievement

What are people trying to do with qubits, anyway? In practical terms, progress with qubits and the quantum computers that they power can be divided into three levels:

- **Computational quantum advantage:** Solving an arbitrary, and almost certainly impractical, task much faster than today's classical computers can manage. Also called *quantum supremacy*. Although controversy continues, this milestone seems to have been achieved. Google first claimed to have reached it in 2019, and then a Chinese team went further with a similar problem in 2021. A special-purpose photonics device has also reached this level.
- **Focused quantum advantage:** Solving a specific real-world problem better, faster, or cheaper than today's classical computers can achieve. (Achievements of this type often set off a game of leapfrog, with quantum and classical computers outdoing each other on a given problem, to everyone's eventual benefit.) Also referred to as *narrow quantum advantage*. Achievements at this level have not happened yet, but we believe that they may appear in the next few years.
- **Broad quantum advantage:** Solving problems that classical computers are not solving today and will never be able to solve, or consistently outperforming classical computers on current problems by one or more orders of magnitude. At this point, we'll see new drugs, batteries, airplane designs, climate models, or logistics efficiencies that could only have been created with quantum computing.

The first milestone, computational quantum advantage, seems to have already been achieved. We like the way Devin Coldewey of TechCrunch described this in a recent article (2022): "The point of quantum supremacy is to show the method's viability by finding even one highly specific and weird task that it can do better than even the fastest supercomputer. Because that gets the quantum foot in the door to expand that library of tasks."

And in mid-2023, IBM claimed to have achieved something they call quantum utility, which is either the same as, or very close to, what we refer to as focused, or narrow, quantum advantage. See Chapter 6 for a brief description. Only time will tell if this claim is later exceeded by classical computers, as IBM has acknowledged.

Winning the Race for Focused Quantum Advantage

We see a lot of emphasis on computational quantum advantage — again, *quantum supremacy* — and on broad quantum advantage, the long-term vision of a world where quantum computers are part of almost everything people do. But we think the intermediate goal — focused quantum advantage — is both critically important and quite near.

As mentioned, this is where quantum computing begins to deliver real advances in solving scientific and practical problems. These advances may occur in two broad areas:

- **Things:** Quantum computing is likely to help to create new drugs, better batteries, new materials, and other things that will be both amazing and worth a lot of money. Imagine a carmaker shipping an electric vehicle (EV) with twice the range of current models, at no increase in cost. Or a new drug that dramatically improves patient outcomes for a specific type of cancer. (One of us authors, Smith, has gone through today's standard treatments for a serious case of cancer, and would have appreciated that.) It seems likely to us that a few amazing things, developed with the help of quantum computing, will show up within a decade.
- **Processes:** Quantum computing is likely to help with weather forecasts, logistics planning, stock trading, assembly line improvements and other processes that will deliver impressively improved results and will, as with the things, be worth a lot of money. Imagine an airline that can cut its costs by 20 percent through improved routing or a weather forecaster charging 20 percent more for more accurate forecasts. These improved processes may not strike the public as being quite as amazing as shiny new objects but will also be worth a lot of money.

We believe that the achievement of focused quantum advantage in just a few products and processes will cause a sudden and very large upswing in interest and investment in quantum computing.

People in various areas of endeavor are often driven primarily by either hope or fear. Much of the interest in quantum computing up to this point has been driven by fear — the fear of conventional data encryption being cracked. But when a few focused uses of quantum computing deliver practical, real-world, and probably highly profitable advantages, hope will predominate. Concerns about encryption will hardly go away, but they'll be less important to quantum computing as a whole than the desire to create and do cool new things and processes with this new technology.

TECHNICAL STUFF

Improvements in AI and ML are a special case of the "processes" kind of improvement. Quantum computing is likely to enable more efficient AI and ML algorithms than is currently possible. Quantum powered AI and ML may take the world to broad quantum advantage — where quantum computing is helping to change our world in large and unpredictable ways — quickly. We're so excited about this that one of us (whurley) even gave a talk about it at the South by Southwest conference in Austin, Texas early in 2023 (see Figure 10-2).

TECHNICAL STUFF

Improvements in things tend to come out of the use of quantum computers for simulating quantum mechanical interactions, as used in designing drugs or more effective batteries. For these applications, the accuracy of each qubit is critical. These applications tend to map well to qubits with lower error rates and higher connectivity, which at this point are seen strongly in trapped ion qubits. Improvements in processes, such as improvements in weather forecasting and AI and ML, come from more purely computational uses of qubits. At present, these applications tend to map well to larger qubit counts, as seen with superconducting qubits.

FIGURE 10-2: whurley's talk on QuantumanAI describes how quantum computing and AI are beginning to be used together.

MARKETING BLATHER QUANTUM ADVANTAGE

We left out a type of quantum computing advantage that is already, unfortunately, being achieved today with regularity: marketing blather quantum advantage. Press releases about venture capital funding, announcements of government investments, and even steps forward in the relevant physics that are actually rather minor fall into this category. These announcements may drive beneficial progress in the field and enhance the reputations of those involved, and for businesses they may drive improvements in hiring, retention, reputation, and even revenue. But all this is the result of announcements that are thin on content, not (yet) solid real-world effects.

Visiting the Qubit Zoo

Most people not directly involved in the industry tend to think that qubits are standardized, just like the bits in classical computers, but this is far from the truth. One of the authors (whurley) travels the world meeting with all kinds of people involved in quantum computing. In conversation with quantum computing insiders, he finds that even those who are upbeat about future progress tend to marvel at the range of technologies under development and the large amount of work that's still to be done.

It's time to introduce the wide and wild range of qubit types that are attracting hundreds of millions of dollars in investment — that's investment per company, with dozens of companies — and vast amounts of effort and interest.

Figure 10-3 compares the six major types of quantum computing modalities. In this case, *major* means a modality that has a reasonable amount of solid theoretical work behind it, is in use by at least one well-funded or public company, and has demonstrated some kind of success in tests.

Qubit development is advancing rapidly, so this figure is a snapshot in time (mid-2023). Expect new developments in this field, as well as continuing controversy as to how one qubit type compares to another.

Deconstructing qubits

Like the term *bit*, the term *qubit* means two different things. For qubits, these two meanings are: a logical element that can hold an infinite range of values, due to superposition, but return only a 0 or 1 as a result; and a quantum electromechanical device that forms the core of a quantum computer.

Quantum Computing Modalities

	Annealing	Universal				
	Superconducting	Trapped Ions	Superconducting	Photonic	Cold/Neutral Atoms	Silicon Spin
Description	Optimization only	Leads in fidelity	Leads in qubit count	May scale rapidly	May scale rapidly	Should be stable
Connectivity	Adjacent-only	All-to-all	Adjacent-only	All-to-all	All-to-all	Adjacent-only
Fidelities	High	Highest (99.9%)	High (99.6%)	Promising (<99%)	Promising (<99%)	Promising (<99%)
Coherence Times	Very short	Seconds	Very short	Long	Seconds	~1 second
Gate Speeds	Fast	Slow	Fast	Fast	Slower	Fast
Qubit Count	Low thousands	Dozens	Low hundreds	Few	Few	Few
Controls	Microwaves and lasers	Lasers and magnetic fields	Microwaves and lasers	Various	Lasers	Lasers and magnetic fields
Extreme Cryo Required?	Yes	No	Yes	No	No	Yes
High Vacuum Required?	No	Yes	No	Yes?	Yes	No
Ability to Manufacture	High	TBD	High	TBD	TBD	TBD

FIGURE 10-3: Comparing modalities for major qubit types.

Qubits are built around coherent matter used for quantum computation. When coherent, they can be placed in a state of *superposition* (being in multiple states at once), can be *entangled* (involved with, but not controlled by, one or more other qubits), and can *tunnel* (suddenly cross a barrier).

The qubit is part of a quantum processing unit that includes circuitry to manage the coherent matter at the heart of the device, but the term *qubit* tends to be used to describe the whole thing. A quantum processing unit requires several elements to work:

- **Coherent matter:** This term refers to a piece of matter that can be kept coherent yet still be influenced as to the value of one or more properties (such as spin), placed into entanglement, and allowed to tunnel from one location to another.
- **Isolation mechanisms:** The coherent matter must be managed by mechanisms that keep it from interacting directly with its environment. Interference such as vibration, heat, and magnetic fields must be kept from decohering the qubit.

» **Control mechanisms:** Coherent matter can be influenced as to properties such as spin, can be entangled, and can be allowed to tunnel without direct interaction that would cause it to decohere. Very subtle influences — from magnetic fields or laser beams, for instance — serve as control mechanisms.

» **Measurement mechanisms:** When a quantum computing program has run, the end state of the quantum computer's qubits must be measured. This measurement yields a 0 or a 1 as a result (the *bit*, short for *binary digit*, in *qubit*) and decoheres the qubit, which must be reinitialized, using the control mechanisms, before another program can be run.

» **Communication mechanisms:** The qubit needs to receive commands from, and return results to, the overall computer of which it's a part. So the qubit needs to communicate in a way that doesn't interfere with the coherent matter at its core.

When assessing the effect of a qubit's error rate on the operation of a quantum computer, we need to keep in mind that cumulative probabilities are multiplicative. If an operation works correctly 90 percent of the time, two operations in a row will operate correctly 81 percent (90 percent x 90 percent) of the time. Adding a third operation means multiplying by 90 percent again, yielding a 72 percent probability of a correct result. Six operations in a row drops the likelihood of a correct result below 50 percent, meaning you have to run a six-step process several times before you can choose a (probably) correct result. (The correct result will recur most often as you do successive runs.) By contrast, a 99 percent accurate qubit that performs three operations will yield a correct result 97 percent of the time; you can perform nearly 70 operations in a row before the likelihood of a correct result drops below 50 percent. Reviewing the cumulative effect of error rates helps to make clear the effect that higher or lower fidelity qubits have on the correctness of results.

Coherence is the necessary condition for matter to display the quantum mechanical properties needed for quantum computing, quantum sensing, or quantum metrology. Matter can be coherent when it's not being measured nor interacting strongly with other matter, and being bound in an atom or molecule is enough to cause the particle to lose coherence. So too is interacting directly with scientists and laboratory or computing equipment. In the wild — when people and their equipment are not messing with them — individual subatomic particles, such as photons, electrons, and ionized atoms, are usually coherent; neutral atoms and molecules are usually not. However, supercooling certain kinds of matter to near absolute zero can put them in a state of coherence. For instance, certain gases form a coherent Bose-Einstein condensate (BEC) when supercooled — and yes, experiments have been done using BECs as qubits. Perhaps someday a computer called BECky, rather than HAL (as in the movie 2001, *A Space Odyssey*), will reveal the secrets of creation to some lucky astronaut.

TECHNICAL STUFF

TOO COOL FOR SCHOOL?

Some qubits — superconducting qubits, which use tiny metal loops, and silicon spin qubits, with individual electrons at their core — require supercooling, which is achieved through *cryogenics*, the science and practice of maintaining extremely low temperatures. *Extremely low* means just above absolute zero, which is –273C, also expressed as 0K, or zero degrees Kelvin. (A degree Kelvin is the same as a degree Celsius, but you start counting from absolute zero and go up.)

As cold as this is, there are still matters of degree (ha ha) when it comes to cryogenics and quantum computing; some supercooled qubits require temperatures in the very low millikelvins (thousandths of a degree Kelvin), while others allow tenths of a degree Kelvin, or even a full degree Kelvin or more, which is much less expensive to achieve and maintain.

Confusingly, all qubits benefit from very low unvarying temperatures, on the order of 1K to 5K, because this reduces thermal interference; but only superconducting qubits and silicon spin qubits require extremely low temperatures, in the millikelvin range.

Cryogenics practice is so advanced that keeping the qubits themselves in the millikelvin range is not prohibitively difficult or expensive. But the control cables that communicate with the outside world are problematic in that they can also transmit thermal interference and other forms of noise. The more cables, the more potential interference, making supercooled quantum computers harder to scale.

Defining quantum computing modalities

A quantum computer's modality is determined by the type of coherent matter at the core of its qubits. Each type of coherent matter used for quantum computing has characteristics that require different approaches to maintaining coherence, controlling the state of the qubit, and communicating with the qubit. So the modalities are named for the type of coherent matter at their core, which affects almost everything else about them: how much isolation they need (including supercooling or no, and the degree of supercooling), how they're controlled, how measurements are taken, and how each qubit communicates with the overall quantum computer.

Many important qualities determine the usefulness of a quantum computing modality for current and future quantum computers. Here, we focus on six:

- **Fidelity:** How close is the final state of the qubit (okay, the coherent matter at the core of the qubit) to the ideal state that a perfect qubit would be in, after the same initialization process and running the same gates? Because error rates compound across qubits, having high fidelities — well above 99 percent — for each qubit is important.
- **Coherence time:** Using (relatively) mainstream supporting technologies, how long can the qubit be kept coherent? The longer that is, the more operations can be accomplished. Coherence times range from a fraction of a thousandth of a second to several seconds.
- **Gate speed:** How long does it take for the qubit to complete a single processing step, or *gate*? Faster gate speeds are better, and gate speed is related to coherence time: A qubit with a short coherence time must have a very fast gate speed, or you'll never be able to run a program with many gates (steps) in it.
- **Scale:** How many qubits can be created and used together in a quantum computer of a given modality (while maintaining isolation, control, measurement, and communication mechanisms)? The more qubits, the more room for error correction and the greater the capability to solve useful problems. In the future, how easy will it be to scale to thousands of qubits?
- **Manufacturability:** Is there a way to manufacture processing units of a given modality in volume and at reasonable cost, while maintaining the required mechanisms and meeting any special requirements, such as supercooling?
- **Entangleability:** Ideally, all the qubits in a quantum computer can be entangled with each other, allowing the system to achieve the maximum possible power from its qubits. Some qubit modalities are easier than others to configure in ways that allow for greater entangleability. Although this capability is often not included with the others in this list, entangleability is of great importance.

Error correction depends on several factors. Current plans are to devote many physical qubits — possibly dozens or hundreds — to creating a single, error-free virtual qubit. The greater the fidelity of each physical qubit, the fewer will be needed to create a virtual qubit. Errors can be reduced also in a top-down manner by using statistical control systems and so on. Progress in error correction currently lags behind progress in qubits.

THE LOGIC GATE REQUIREMENT

Quantum computing uses logic gates that require either one or two qubits each to implement. If qubits operate at 95 percent reliability, a one-qubit logic gate will perform the right operation — you guessed it — 95 percent of the time. But a logic gate that requires two qubits will operate correctly only about 90 percent of the time, which means that running several one- and two-qubit logic gates in a row becomes unlikely to yield a correct result.

Quantum annealers, being optimization rather than calculation machines, are more forgiving. As annealers improve, errors will cause them to deliver a result that's imperfect, but it may still be better than anything affordably available from classical computers. And if your daily delivery dispatching operation is 10 percent better than the competition, that can be pretty valuable, even if you're still 5 percent short of perfection.

Mapping Out the Modalities Landscape

Among the factors that drive qubit development forward, two stand out:

- **Fidelity:** The degree to which a qubit's state approximates that of an ideal qubit, from the beginning of a program run to the end. (Degradation at any step is amplified by entanglement amongst qubits.) Simulation, as used in materials research and drug discovery, is the most demanding type of application where fidelity is concerned.
- **Scale:** The current qubit count for quantum computers of different types. Computation, as used for Shor's or Grover's algorithm (see Chapter 6) or for AI and machine learning, is the most demanding type of application where scale is concerned.

Figure 10-4 shows a very rough map of how different quantum computing modalities currently compare as to fidelity and scalability. Your mileage may vary, and any modality may show rapid progress at any time.

REMEMBER

The quantum computing landscape is so complex that, in this chapter, we're directly discussing different types of qubits as used for only gate-based quantum computing. However, you may end up using a quantum computer simulator running on classical hardware for learning steps or for optimization problems (Chapter 8). You may also end up using a quantum annealer (Chapter 9). All types of quantum computers are part of the same overall learning curve, and you need to stay aware of the entire scene as you proceed. Experience in any one area of quantum computing will help you in all areas when needed.

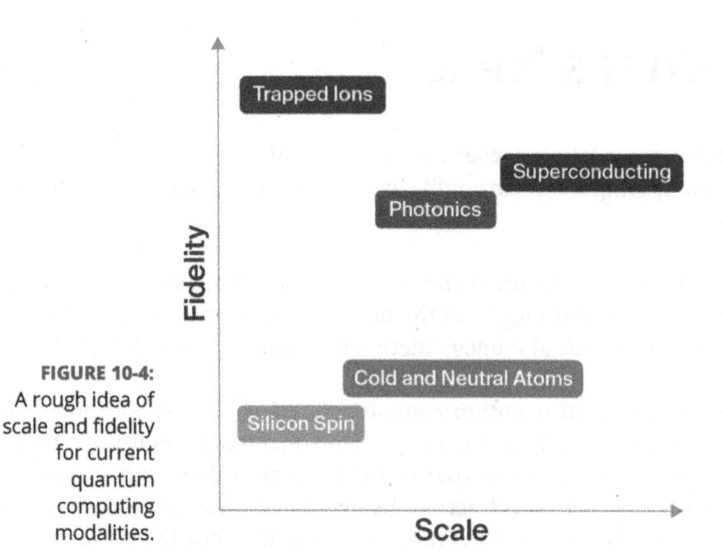

FIGURE 10-4: A rough idea of scale and fidelity for current quantum computing modalities.

THIS ERA'S MOONSHOT

People working on different quantum computing modalities frequently describe the difficulty of creating and corralling qubits to the effort to put people on the moon more than 50 years ago. Like the moon landing, creating useful qubits requires experts from many fields. While physics is at the forefront, pure and applied mathematics, materials science, chemistry, electrical engineering, and other fields all play an important role. (Ironically, quantum computers, once scaled up, hold promise for helping in all these fields.)

We don't yet have a quantum computer that solves real problems better than existing devices. Electronic computers reached this level with the codebreaking work done by Colossus in World War II. (See Chapter 2.) From Colossus, it took 10 years before vacuum tubes were replaced with transistors, and 20 more years before the microprocessor was created. Today, control of the technology used to make the most advanced microprocessors is a matter of global concern. (For more information on the critical economic, political, and military effects of microprocessor technology, check out *Chip War* by Chris Miller [Scribner, 2022].)

These time frames give us some idea of how long it might take for qubits to become fully useful and to become standardized. And the importance of microprocessor technology today shows how important getting qubit technology right may become.

Figuring Out What's Next

In the next chapter, we take a deeper dive into each of the major types of qubits. The quantum computing zoo seems unlikely to cohere around a specific qubit type anytime soon.

Earlier in this chapter, we mention that we expect a half dozen or so quantum computing-powered breakthroughs in the not-too-distant future, and that such a breakout will cause a rush of money, talent, and attention toward the field.

It's still very early days in quantum computing, and we are at very low qubit counts, lacking error correction, missing out on manufacturability — there's much more work to do than work that has already been done. But if quantum computing bags some real-world accomplishments in these early days, it will appear (correctly, we believe) that the future of the field is nearly limitless.

Investors will be eager for a piece of that future. So we expect that a surge of early successes will cause more money and effort, not less, to be invested across the whole range of qubit types.

A virtuous circle is also on the horizon, much like what we saw for classical computing over many decades. As each generation of classical computers has improved, the next generation has been designed on better and better classical computers, yielding ever-improving results over many decades. It's only now, after 50 years of exponential improvements, that Moore's Law appears to be running out of steam.

This scenario is likely to also happen in quantum computing. Quantum computers from one generation will help to find the optimal designs for the next generation. They will also help to create materials that might, for instance, allow supercooling to become relatively cheap and easy, or remove the need for it entirely.

As Rocky used to say, in the classic *Rocky and Bullwinkle* television series: "But wait! There's more!" Quantum computing is likely to take AI and machine learning to new levels. So we might have a new, AI-powered partner intelligence helping to design the next generation of quantum computers, which will then power even higher levels of AI and machine learning.

At that point, the future of quantum computing will be bright indeed. The future of humanity may become rather uncertain, but whatever the form factor inhabited by the intelligences that end up running things going forward, they will no doubt have some rather awesome quantum computers.

» Finding trapped ions and superconductors and more, oh my!

» Identifying your best strategy for quantum computing

Chapter 11
Choosing a Qubit Type

The Space Race began as a mano-a-mano contest for dominance between two nation-states and their allies. But once the initial target was reached, and humanity had walked, driven around, and hit a few golf balls on the Moon, competition in space went relatively quiet for decades. Only now are there plans to return people to the Moon. (We're guessing we'll see some quantum advantage before more moonwalks happen.)

And a funny thing happened: The Space Race is no longer a one-on-one sprint between two competing political ideologies. It's more like a space decathlon, with lots of competitors, many events, and more than one way to stand out in the crowd.

The quantum computing world is evolving in the same direction. In geopolitical terms, there is a new rivalry, with the West — including the US, Canada, the UK, Australia, and the European countries of the EU — largely competing with China. Meanwhile, a good deal of scientific cooperation still exists, and we hope that cooperation will far outweigh competition going forward.

Currently, China seems to be leading in quantum communication, with many achievements and a huge chunk of patents. The West seems to be leading on quantum computing. However, as we discuss in this chapter, it's still early days; importantly, the field has not even reached quantum advantage yet.

In terms of the choices you have to make, it's a decathlon as well. You'll have a choice among quantum inspired computing on classical computers (Chapter 8), quantum annealing (Chapter 9), and gate-based quantum computing. In the gate-based world (Chapter 10 and this chapter), you'll have your pick of qubit types.

In this chapter, we drill down on the different gate-based quantum computer modalities and how that might map to your goals in quantum computing. You may find yourself making a choice to pursue a given modality for a while or for a specific project. But that choice, like the operations supported by quantum computer logic gates (see Chapter 12), is reversible.

Anyone entering the field of quantum computing should be generally aware of the technological and industry landscape described in this chapter and the preceding chapter. But only those who are committing to serious investment of time, effort, and money into quantum computing need to consider selecting a specific type of quantum computer today. Everyone else can switch among different types of quantum computers that are easily available through the cloud (see Chapter 15).

Quantum computing is still in the NISQ era: noisy, intermediate-scale quantum computing. *Noisy* refers to the fact that each qubit and its control circuitry generates noise that can affect nearby qubits; it also alludes to the vulnerability of each qubit to noise from its neighbors and the overall environment. *Intermediate-scale* refers to the number of qubits involved, though the dozens to hundreds of useful qubits available today might be viewed as chump change in the longer run. As for the *quantum computing* part, we aren't really *doing* quantum computing yet; rather, we're *experimenting* with it. Useful results are coming from quantum-inspired computing running on classical computing hardware (Chapter 8) more than on quantum annealers or gate-based quantum computers themselves.

Telling the Players Apart with a Scorecard

There's an old saying that goes, "You can't tell the players apart without a scorecard." Now this saying comes from the old days, when baseball players had only numbers on their jerseys, not names along with the numbers. (This practice made it cheaper and easier for management to replace players). But it's often used to describe situations in sports, business, and other areas where every player, or option, is solid, but none are outstanding.

There's some truth to this saying in quantum computing. Among the different modalities, there are some early leaders, but you can't take your eye off any of the major options just yet. Figure 11-1 shows the current major players in each quantum computing modality.

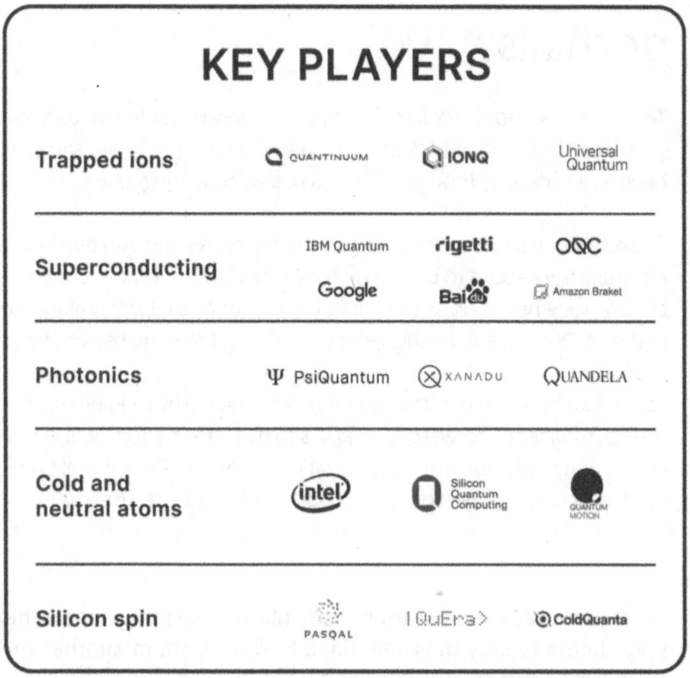

KEY PLAYERS

Trapped ions	QUANTINUUM	IONQ	Universal Quantum
Superconducting	IBM Quantum	rigetti	OQC
	Google	Baidu	Amazon Braket
Photonics	PsiQuantum	XANADU	QUANDELA
Cold and neutral atoms	intel	Silicon Quantum Computing	QUANTUM MOTION
Silicon spin	PASQAL	QuEra	ColdQuanta

FIGURE 11-1: Quantum computing vendors by modality.

There are more types than we show here, such as qubits made from diamond vacancies and trapped electrons. (We give a brief mention to these additional types later in the chapter.) The main point is that you should keep updating your scorecard.

Just why are there so many qubit types? The reasons for the proliferation of qubit types — and why there aren't even more qubit types under commercial development — have as much to do with business considerations as technical ones. Keep this in mind as you choose which vendors to work with in quantum computing.

We have early leaders in superconducting qubits (Google and IBM) and in trapped ion computing (IonQ and Quantinuum). If a founding team and investors want to start a new quantum hardware company today, they have a lot of catching up to do with either of these technologies and must also purchase a lot of equipment up front.

Other approaches — photonics, cold and neutral atoms, and silicon spin — are newer. They may turn out to offer surprising advantages for broad use. Or they may be well-suited to focused problem sets and use cases. An example of a possible niche use is military or industrial use, where hardening against shocks, portability, extra security features, or low cost per unit may be important. A company may only need to make progress in one niche to stay in the game.

QCFDaaS?

Technology vendors love to sell things "as a service": software as a service (SaaS), platform as a service (PaaS), infrastructure as a service (IaaS). But sadly, we can't offer this book as a service — though QCFDaaS has a certain ring to it.

Books about a specific technology — even e-books and audiobooks, delivered electronically — tend to be things. They reflect a snapshot in time. The chapters must all hang together — or, as Benjamin Franklin once said, with reference to the signers of the US Declaration of Independence, they will all hang separately.

You're making a big investment of time and energy by reading this book. But how can you keep current afterward? The key is to develop relationships in the industry. Get to know a range of people and organizations. Keep up. Contribute. Some of the communities and resources mentioned in Chapters 15 and 16 can be useful in this regard.

Even personal considerations can matter. Leading researchers in one modality may choose to stay in academia; a leading team in another modality may want to get funding and start a company.

So don't feel that you have to keep up on specific developments in every quantum computing modality. Just master the basics, as presented here, and then keep an eye on key developments. Important progress in any modality is likely to get a lot of attention and give you the opportunity to revise your approach if needed.

When assessing the claims of different vendors, keep in mind the difference between comparisons within a modality and between modalities. If a vendor says "our approach is very scalable," do they mean compared to all modalities — superconducting qubits are currently the most scalable — or simply among vendors within a specific modality, such as trapped ion systems? Also be aware that trapped ion and superconducting modalities are well-established enough to have a high likelihood of staying power for large classes of problems. At this writing, photonic, cold and neutral atom, and silicon spin approaches have not yet demonstrated this level of achievement.

Qubits require coherent matter — usually a subatomic particle or an atom — that is not interacting, nor being measured directly, at their core. The matter must have two opposed states, for instance, the spin of a subatomic particle such as an electron, which is either up or down. The coherent matter is placed into a state of superposition, such that it is in both and neither state (both and neither spin up and spin down) at the same time. Each individual chunk of coherent matter must be kept isolated from its environment and free from direct measurement, or bad things happen: Either it decoheres entirely or interference prevents it from returning a correct result. But at the same time, the control plane must be able to entangle qubits together and to influence the spin or other property that's in superposition, all without causing decoherence or allowing errors to creep in. These are just the first level of technical requirements that must be met for a qubit to be useful in quantum computing.

Trapped ion qubits

Trapped ion qubits were the first discrete qubit type to be demonstrated in use, back in the 1990s. Trapped ion devices have reached as high as 32 high-quality qubits, in a device from Quantinuum, a spinoff company from computer maker Honeywell. IonQ, the first quantum hardware company to become a public company, is another leading company in this space. A lot of venture capital and other money has been invested in this technology.

As the name implies, trapped ion qubits (see Figure 11-2) are created by ionizing a neutral atom by using lasers to strip away an electron. This leaves the atom with a positive charge. The ionized atom can be held in place with magnetic fields (which work really well with ions), and the ion is kept in a vacuum to avoid interference. The ion can then be placed in superposition and manipulated with additional laser bursts to run gates (perform steps in a quantum computing program).

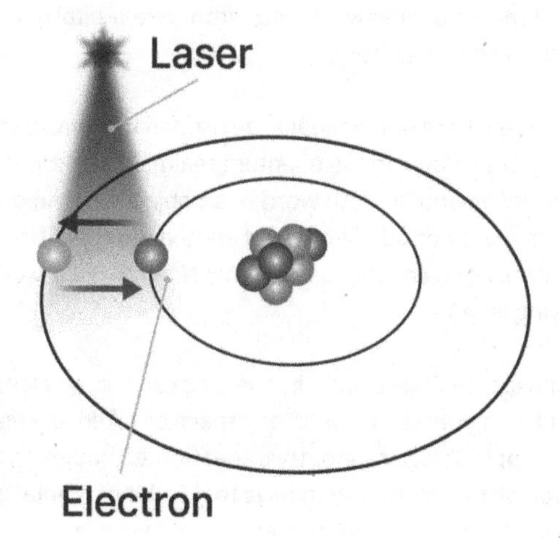

FIGURE 11-2: Ionized atoms are easily controlled by magnets in trapped ion quantum computing.

While supercooling to well below 1K is not required, cooling the trapped ions to 4K makes coherence much easier to maintain. (Consistent temperatures in the 4K range are more easily achieved than fully supercooled temperatures a fraction of a degree above 0K, as required by superconducting qubits.) However, maintaining the hard vacuum required by trapped ions is not easy. Trapped ions are not yet manufacturable in the same volume as superconducting qubits; there are still scientific unknowns as to how, or even whether, trapped ion systems will scale.

The electron is then manipulated by further laser bursts as part of running the quantum computing program. The laser pumps the valence (outermost) electron full of energy so it moves to an outer orbital, far from the nucleus, single and ready to mingle with other overstimulated electrons in other ions. This makes it easy to entangle trapped ion qubits with each other, to a rather risqué degree. (For incredible quantum reasons that we won't explain here, all the ions in a bunch of several dozen can be easily entangled with each other, even if they aren't physically adjacent. And no, your humble authors are not creating an iconic dating app based on this capability.)

WHY QUANTUM LEAPS . . . AREN'T

A core understanding from quantum mechanics is that electrons can be in only specific orbitals (potential locations); the distance between orbitals is dictated by Planck's constant (see Chapter 3). When an electron gains or loses energy, through the absorption or emission of a photon, it moves suddenly, or leaps, from one orbital to another, which is where the term *quantum leap* comes from.

However, this is another area where our language, developed in a macroscopic world ruled by classical physics, lets us down in the quantum realm. In a classical leap — a *salto*, to use the evocative Italian and Spanish word — an object starts in one place, moves through space over some period of time, and ends in another. (Unlike many more hazardous undertakings, you *can* try this at home! Now, or the next time someone tells you to take a flying leap.)

In a quantum leap, in contrast, an object starts in one place and, in no elapsed time, appears in the other, not having passed through any space or taken any time in the process. (You, being a macroscopic object, cannot try this at home, though it's happening around you and within you all the time.) So you have to do a bit of mental gymnastics to understand what really happens in a quantum leap.

Also please note that most professionals in the field *hate* the term *quantum leap*. We cheerfully use it here, however, because we're reaching a wider audience and to discomfit the pros.

This ease of entanglement adds a lot of power to trapped ion quantum computers. The ion is then read by a final laser burst as being in a high energy state (0) or a low energy state (1). All of this is tracked by imaging the absorbed and emitted photons in a highly accurate process.

But the process takes time. Trapped ion quantum computers run programs relatively slowly, but they still achieve high circuit depth (completing many programming steps) due to their long coherence time of many seconds. They also have a great deal of quantum computing power due to their all-to-all entangleability. (Yes, we just made that word up; our spell checker is having a fit.)

Unfortunately, scaling trapped ions is hard, because the vacuum and control machinery gets baroque beyond a few dozen qubits. Also, charged particles (including ionized atoms) strongly repel each other, so forcing them into close proximity is difficult and generates noise.

Makers of trapped ion systems like to measure their capacity using metrics such as quantum volume (Quantinuum) or algorithmic qubits (IonQ). These measurements are derived from inputs such as the number of qubits, their fidelity, and their capability to execute programming steps, rather than simply qubit counts. Using some kind of alternative measurement is reasonable, to capture the complexities involved. For trapped ion devices to be part of the future, however, we need steady progress in the qubit count, even if the count is not as high as (noisier) superconducting qubits.

Superconducting qubits

Superconducting qubits are the clear leaders in scale, with IBM having achieved more than 400 qubits in its newest devices and set to deliver a device with over 1,000 qubits in the near future. Google, IQM, and Rigetti have systems with between 50 and 80 qubits. Oxford Quantum Computers and Chinese company Baidu have much smaller systems with 8 to 11 qubits. Amazon has also announced its intent to build a superconducting quantum computer, though it does not have any functional systems to date.

TECHNICAL STUFF

To make matters more confusing, within modalities there are different types of qubits. Today's most widely used superconducting qubits (see Figure 11-3) are also called *transmons*, which is short for *transmission line shunted plasma oscillation qubit*. (Just remember *transmon*.) Other types of less popular superconducting qubits include the coaxmon and fluxmon.

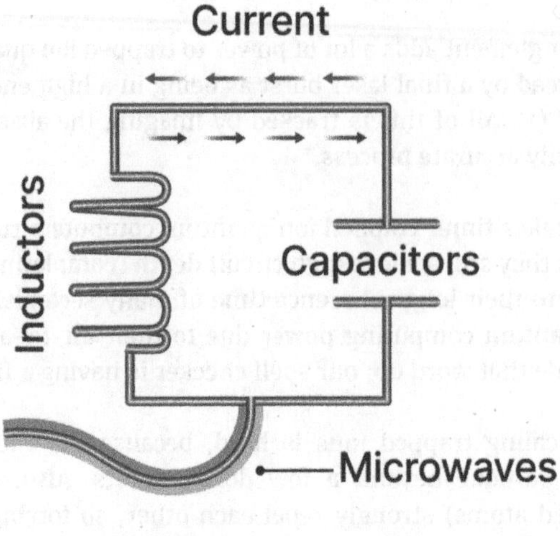

FIGURE 11-3: Superconducting loops are at the core of transmon qubits.

Superconductors are specific types of metals, ceramics, and a few other materials that, when chilled to a very low temperature — called the critical temperature for a given material — carry current without loss. At or below the critical temperature, electrons go from strongly repelling each other (as they do at all other temperatures) to mildly attracting each other; they pair up, and these paired electrons (Cooper pairs) flow freely.

Superconducting qubits are unique among quantum computing modalities in that the coherent matter at the core of each qubit is not a single atom or a subatomic particle, which are often coherent when left to themselves, but a loop of metal that is very small by normal standards but is still made up of billions of atoms. The relative complexity of transmons means they vary from one another and must be calibrated before use.

Supercooling makes each loop capable of coherence. Two of these supercooled loops are placed very close to each other, with a barrier between, such that Cooper pairs tunnel between them. Different properties of the electron pairs are used as the quantum property that's placed in a state of superposition. Superconducting qubits are placed into superposition by microwaves.

These loop/barrier assemblages, exotic as they might seem, can be manufactured with existing circuit printing processes, similar to the process used to manufacture traditional semiconductor integrated circuits. Their relatively large size makes them more manageable than other kinds of qubits. They are referred to as *artificial atoms* because they are manufactured (the *artificial* part), but they are, like an atom, still subject to quantum mechanical principles. (Silicon dots, though much smaller, also act as artificial atoms.) However, the far larger size of the coherent matter in superconducting qubits and their requirement for extreme cryogenics makes them less likely to stay coherent and stable than trapped ion qubits.

Superconducting qubits run *gates* — that is, complete programming steps — very quickly, but their fidelity is low. As a result, superconducting qubits run only a limited number of gates before they're likely to fail. (This is called "having low circuit depth" because the qubit only gets so deep into a series of gates before it fails.) This tendency toward decoherence increases the need for error correction in superconducting versus other types of qubits.

Superconducting qubits are well-suited for creating one-qubit and two-qubit logic gates, which makes them suitable for calculations. IBM is blazing a trail for the quantum computing world with its regularly updated roadmap (see Chapter 1) for future quantum computers with ever-higher qubit counts, doubling in number every one to two years in a very Moore's Law–type way, and it is doing this with transmon qubits.

IBM has consistently met their quantum computing roadmap goals first laid out in 2020 and is setting such a rapid pace that other qubit types, even if they have higher fidelities and greater theoretical capability, may be challenged to keep up.

D-Wave is also blazing a trail with its quantum annealers, which have several thousand superconducting qubits, compared to the several hundred in current gate-based quantum computers from IBM. To achieve this greater qubit count, D-Wave integrated control circuitry onto qubits and accepted limited entanglement among qubits. These design choices are acceptable for an annealing approach, which is somewhat less vulnerable to errors than a logic-gate approach. (A noisy annealer might produce a result that, while not perfect, is still useful; a noisy logic-gate computer is likely to be either wonderfully right or seriously wrong.)

Some superconductors operate at much higher temperatures than others, with the highest so far being around 133K — almost halfway between 0K (-273C) and 0C, the freezing point of water (for our fellow Americans). However, these high-temperature superconductors, as they are called, are not suitable for use in qubits. Future quantum computers, being so well-suited for materials research, may be able to help create materials that both exhibit superconductivity at higher temperatures and can be used in qubits (and, we hope, starships). The pairing of AI and quantum computing, already underway, may also be useful in this search. The dream is of a room-temperature superconductor that can be used for qubits; every substantial step of progress in this direction is potentially useful.

THE AGONY AND THE ECSTASY OF TODAY'S QUBITS

Superconducting qubits and trapped ion qubits are, unfortunately, almost logical complements when it comes to desirable properties for quantum computing.

Superconducting qubits have lower gate fidelity across an entire quantum processing chip (median fidelities around 97 to 99 percent), have very brief coherence times (a millionth of a second), can be connected only to adjacent qubits, and must be cooled to absolute zero. These qubits are "unnatural" because they are manufactured. They are complex (being made up of billions of atoms each) and therefore variable, so each qubit must be tuned and retuned for use.

However, the superconducting qubit is very fast, relatively easy to manufacture (partly because it's larger), and easy to connect to adjacent qubits. It's also scalable; IBM has multiple systems with more than 100 qubits (if only they weren't so error-prone). The needed breakthrough is a combination of noise reduction per qubit, with fidelities consistently exceeding 99 percent, and error correction across qubits.

A trapped ion qubit has much higher fidelity (averaging about 99 to 99.8 percent at present) and much greater connectivity (all the qubits in a row of qubits can be entangled with each other pairwise). Gate times are slower, but this is offset by the qubits staying coherent for many entire seconds. Being natural and simple at their core, the qubits are relatively invariant and do not need tuning.

Because the coherent matter at the core of it is so small, the entire qubit can also be very small, but that makes it hard to manufacture with current technology. This makes it hard to increase the qubit count. (Let's put our million-qubit quantum computers and their AI programs on that right away — oh, wait . . .)

With the atom at the core of each qubit being controlled by tiny lasers, the assemblage doesn't scale well beyond a few dozen qubits. Trapped ion quantum computers have recently moved up from around 10 qubits to their current peak of 32 (very reliable) qubits, and exciting advances in materials research are promised whenever they can reach 80 qubits or more. The needed breakthrough here is a combination of better control technology — laser tag stops being fun at the atomic scale — and manufacturability.

We may see both trapped ion and transmon qubits types progress steadily or one break out and race ahead — if one of the other qubit types doesn't beat both currently leading types to the punch. It's still all to play for.

Photonic qubits

Once you start thinking a lot about qubits (to quote the late, great US Senator and presidential candidate Bob Dole: "Just don't do it"), the appeal of building qubits around photons is clear. Tiny, massless, and chargeless, a single photon can be used for multiple qubits by manipulating different quantum properties, such as its polarization, the path it follows in a beam splitter, or whether it arrives at a specified location early or late. In principle, you could use two or more properties per photon, easily scaling your qubit count.

Photonic qubits (see Figure 11-4) have a very long coherence time. They don't interact much with the environment — they're too busy being massless and chargeless and moving at lightspeed — are easy to entangle with each other, and can be manipulated in a vast number of ways. However, because they want nothing to do with one another, they're hard to combine into two-qubit logic gates. Some of the equipment used with these qubits requires supercooling, but the hope is to remove that requirement in the future. (The photons themselves deeply do not care about your Earth temperatures.) PsiQuantum, Quandela, and Xanadu are leading companies pursuing photonic qubits.

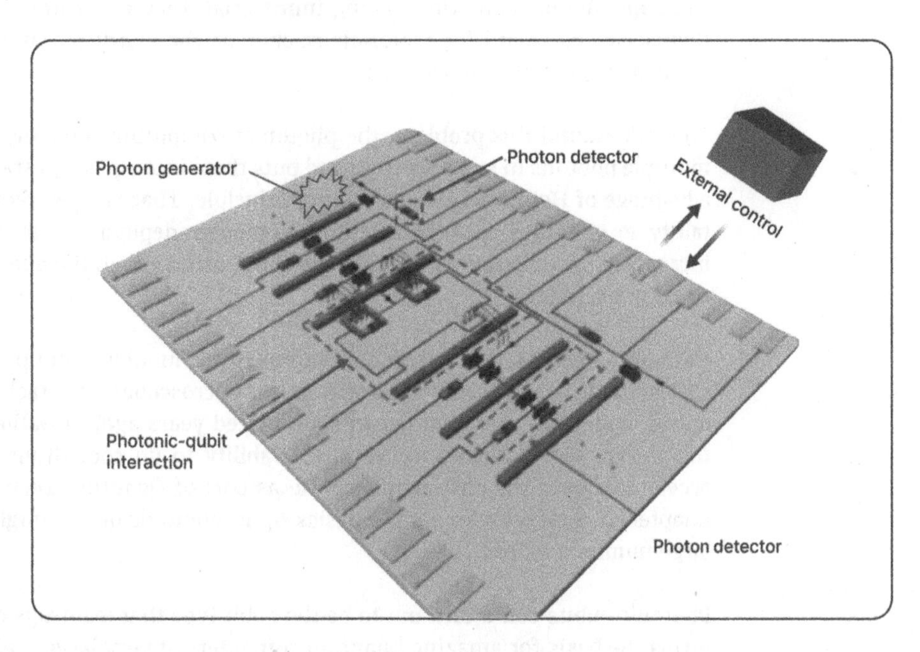

FIGURE 11-4: Photons can be used for qubits in a variety of ways.

TECHNICAL STUFF

We call photons *objects* because we have to call them something, but a photon is simply a bundle of energy; it has no mass, by definition, and no deterministic location, only a probabilistic one. As Gertrude Stein once said of a city she disliked, "there's no there there." Additionally, the centroid of a photon's probability cloud moves only at the speed of light, whereas no massive object can move that fast or even close to it. (If an object tries, it stops aging and then expands to approach the size of the known universe. No jokes about our dieting failures at this point, please.)

While photonic qubits are a natural choice for specific information-processing tasks and for quantum communication — did we mention that they can move only at the speed of light? — they're hard to corral into larger numbers of coherent qubits for quantum computing tasks. While there are many ways you *can* manipulate photons en masse to do quantum computing, the "happy path" as to how you *should* manipulate photons en masse to do quantum computing is not yet clear.

Every few months, a lab somewhere will report a truly amazing new accomplishment with individual photonic qubits (as well as with other kinds of qubits), but the truly tiny and extremely fast little buggers are hard to manage. So far, it hasn't been possible to round them up into reliably forming useful numbers of manageable logic gates, especially the entangled two-qubit ones. (Photons are thoroughly uninterested in one another and in you.) This makes it hard to use them for calculations. At the same time, being immaterial, they don't immediately present themselves as useful for materials science or chemistry research, as can more easily be done with trapped ions.

To work around this problem, the photonics computing company Xanadu places multiple photons in superposition and puts them into squeezed states, which take advantage of Heisenberg's uncertainty principle. That is, they reduce the uncertainty in variables that the quantum computer depends on at the expense of increasing the uncertainty in other variables, about which the quantum computer could not care less.

Humanity does have considerable equipment for manipulating photons at larger scales due to all the work with telescopes, microscopes, spectacles, lasers, fiber optics, and other uses that began six hundred years ago, in Galileo's time (now there's an R&D pathway for you). Our ability to do cool things with photons accelerated over the past century or so as part of Quantum Technology 1.0 (see Chapter 4). So if we ever get the basics right, photonic qubits might scale to very large numbers indeed.

Photonic qubits could turn out to be the qubit type that leapfrogs other work and forms the basis for amazing quantum computers at very large scale in just a few years. However, they may instead perform amazing tasks in quantum communications and quantum metrology (measurement) but not serve as the basis for the quantum computers of the future.

WHEN SOPHONS ARE NOT SO FUN AFTER ALL

One of your humble authors (Smith) is a big fan of *The Three-Body Problem*, a best-selling science fiction trilogy by China's leading science fiction author, Liu Cixin. "Three-body," as fans call it, should be out as a series on Netflix by the time you read this. In the books (mild spoiler alert), sophons — tiny supercomputers, each encapsulated in a single proton — travel from, well, elsewhere to Earth. Once arrived, they serve as spies, and other desiderata, advancing the cause of their masters. One of them even takes human form. Sophons are just one of the ideas that appear in the trilogy that seem to be inspired by Quantum Technology 2.0 (or 10.0 or something).

The best of the rest

"I am large, I contain multitudes," wrote the poet Walt Whitman in 1855. We're pretty sure he wasn't referring to the number of different types of qubits that would be under development in this period, but perhaps he was. We mentioned that he was a poet?

Here are brief descriptions of a few other types of qubits that are still more in the "research" than the "development" stage of the R&D cycle:

- **Neutral/cold atoms:** Similar to trapped ions, neutral (aka cold) atoms are trapped by laser beams rather than magnetic fields. Single outer-shell electrons are excited by lasers to do the quantum part. These qubits are easily entangled and, unlike ions (which naturally repel each other), can be packed tightly into an array, forming powerful multi-qubit gates. Infleqtion, Pasqal, and QuEra use this technology.
- **Silicon spin:** Silicon spin qubits are conceptually similar to transmon qubits, in that they're also artificial atoms. But instead of chunks of superconducting metal, they use a single atom of a semiconducting element (such as silicon). A single electron, controlled by microwaves, is the quantum mechanical actor. The qubit may be placed in a quantum dot, enhancing manufacturability because lots of quantum dots are already out there, such as in every pixel of many TV screens. This type of qubit is backed by Intel, Quantum Motion, and Silicon Quantum Computing. As the Intel connection might suggest, silicon spin qubits are meant to be highly manufacturable using chip technology, but error rates are currently high.

WHEN YOU'RE HOT YOU'RE HOT

. . . and when you're not, you're not. Most of quantum computing operates at millikelvin temperatures, mere thousandths of a degree above absolute zero. These days, temperatures this low are not *that* difficult for technicians to achieve and maintain. But communicating between millikelvin world and room temperature world requires physical connections that introduce thermal noise into the cold place, disturbing the coherence of supercooled qubits.

Two recent developments advance the state of the art a bit. The first is "hot" silicon spin qubits from Intel and QuTech. These qubits operate at an entire degree Kelvin! Believe it or not, this is a significant step toward improving qubit fidelity and reducing the cost and environmental effect of quantum computing.

The second is a new control chip from quantum computing startup Seeqc that has qubit control circuitry, not actual qubits. The chip is designed to operate at millikelvin temperatures and to be bonded directly to qubits, simplifying the task of managing qubits that require extremely cold temperatures. (You need just one connection from room temperature world to the new chip, which lives in millikelvin world along with the qubits, to control a bunch of qubits. This eliminates the need for one or more quantum mechanically "noisy" wires per qubit.)

These advances are examples of the kind of slow but steady progress that doesn't make big headlines but will sooner or later get us the kind of real-world advances that the entire quantum computing effort is anticipating.

A *quantum dot* is a semiconductor nanoparticle that glows a particular color when illuminated. It's used in a wide variety of applications, including medicine and solar power, as well as in display screens. One of the authors (Smith) remembers watching the beginnings of the Space Race on a black-and-white television and seeing color television for the first time — *Rocky and Bullwinkle*, of course — so for him to now be writing about qubits captures a fairly broad range of technological advance.

Quantum dots could theoretically be used as qubits directly, either singly or in pairs; silicon spin qubits can also be inserted in quantum dots. So if you hear of a system using quantum dot qubits, ask whether it's a silicon spin qubit or not.

There are claims that Majorana fermions, which would be well-suited for error correction, have been observed within the last decade, but these are the subject of controversy. And sighting them, once confirmed, is still likely to be a long way from using them as qubits.

There are also quantum computer modalities that are currently minor, in investment terms, which could become major at any time. Here are two examples:

- **NMR (in vitro) qubits:** Matter kept as a liquid in a test tube can be used for quantum computing by picking out a few atoms or molecules and using their spins as the basis for qubits. (*In vitro* is Latin for *within the glass* — the glass, in this case, being the glass of a test tube.) This is how the first quantum computing lab demonstrations were achieved in the 1990s (see Chapter 6 and Figure 11-5). They are not currently being used in commercially funded work.
- **BEC qubits:** Qubits can be made from superconducting gas in the form of a Bose-Einstein condensate (BEC). This has been done experimentally, but not yet taken up for commercial use. BECs would be somewhat similar to superconducting qubits made from metal, in that each qubit would be made up of a huge numbers of atoms. To us, they just sound really cool. (And yes, they would require cooling to the millikelvin range.)

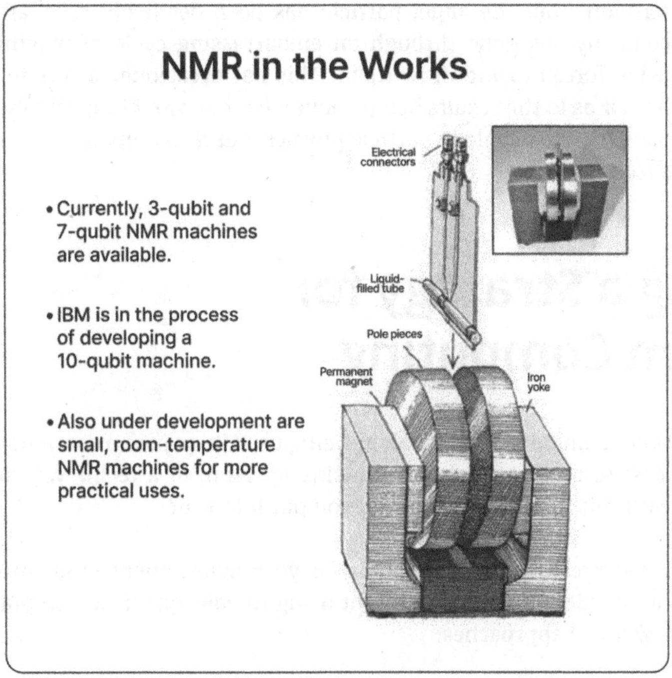

NMR in the Works

- Currently, 3-qubit and 7-qubit NMR machines are available.
- IBM is in the process of developing a 10-qubit machine.
- Also under development are small, room-temperature NMR machines for more practical uses.

Electrical connectors
Liquid-filled tube
Pole pieces
Permanent magnet
Iron yoke

FIGURE 11-5: NMR qubits were vital in bringing quantum computing to life in the 1990s.

In addition, certain qubit approaches could be used with different physical implementations than those just listed, or with, say, superconducting or trapped ion qubits:

- **Topological qubits:** A hypothesized particle called a Majorana fermion — referred to as the *angel particle* — would have outstanding error-correction qualities. Important steps have been taken toward making it work, once the required angelic visitation has occurred. Recent progress has been made in implementing a topological approach atop trapped ion qubits from Quantinuum.
- **Defect-based qubits:** Invisible flaws in a diamond can be controlled with good fidelity at room temperature. Successfully using these flaws to form qubits could make diamonds everyone's best friend. Other kinds of flaws, such as a missing qubit in an array of other kinds of qubits, can also be used as defects as part of this approach.

TECHNICAL STUFF

Microsoft has based their quantum computing hardware strategy on Majorana fermions, but the angel particle has been devilishly reluctant to manifest. The company has gone through an embarrassing cycle of making claims and then being forced to withdraw them. This has occasioned a certain amount of unkind humor as to the results being about what one would expect when a software company tries to dabble in particle physics, but the company's efforts may yet achieve success.

Choosing a Strategy for Quantum Computing

Understanding the technology being used in quantum computing is fascinating, at least to us. But what can a businessperson, or a techie working in business, do with this knowledge? How do you put it to use?

We suggest that you set goals for your involvement in quantum computing. We can divide the goals you might bring to your quantum computing journey into a few broad approaches:

- **Learning and strategizing:** We hope that this book is part of the learning curve for many people. Once you have a strong understanding of the quantum computing landscape, players, and applications, you may want to work with your organization to put together a strategic plan, as described in Chapter 8.

- **Training:** If you're technically oriented, you may want to attend a training course in quantum computer programming (see Chapter 12). You can learn how to program a quantum computer, try a range of different algorithms, and perhaps experiment with applying a few of them to opportunities and challenges you see in your business. If you're not technically oriented yourself, connect with people who are.
- **Opportunity assessment:** Many tech-savvy organizations are creating internal teams that spend significant time using several different kinds of quantum computers to find opportunities for competitive advantage (or threats of competitive disadvantage, or both), now and in the near term. For this approach, you're helping the organization map the current state of the industry and major vendors' offerings to the likely near-term and longer-term needs of your current business, new opportunities, and competitive threats. You'll start to find likely partners in your future quantum computing journey.
- **Project work:** A few organizations are committing serious resources to solving one or two truly important business challenges using quantum computing. These organizations are choosing the right type of quantum computing for the challenge(s) they're addressing and engaging with vendors and consultants as they begin serious work. This approach has the highest risk — the vendor you choose today may not be the one you end up with tomorrow, same same for the vendors and consultants — but also highest potential rewards. (The words *same same* here are not a typo; this is an expression used in aviation to overcome the effects of static on spoken communication.)

Taking advantage of the cloud

One game-changing new feature of the computing landscape gives quantum computing vendors and customers a tremendous advantage: the advent of the cloud. The cloud gives you two advantages in engaging with quantum computing, one obvious and the other more subtle.

The obvious advantage is flexibility. You need to make only a very small investment to start working with quantum computers, because you pay for minutes of use, without having to buy your own quantum computer just to get started. There are currently a lot of free credits are floating around out there, which makes getting started even easier.

On the vendor side, quantum computer vendors can engage with new customers relatively easily. At this writing, quite a few quantum computing vendors are chasing a relatively small amount of spending. It's a buyer's market, at least as long as you're "trying" quantum computing rather than seriously "buying" time on advanced quantum computers, where the costs can really start to add up.

CLARIFYING QUESTIONS

If you're considering finishing this book and then setting quantum computing aside for a while, we urge you to consider a few clarifying questions.

At the individual level, AI and machine learning are improving to the point that they will augment some knowledge workers' jobs and replace others. At the same time, many techies are being pushed out of pure-play tech companies and into more traditional organizations that may be more stable but offer less technical challenge and lower compensation. If you want to stay in the tech industry, putting focused time and effort into quantum computing might get you ahead of the crowd.

At the organizational level, quantum computing, broadly considered, is starting to deliver real business value. It's giving companies an opportunity to look hard at their compute-bound problems and examine whether quantum computing can help. In some cases, these companies are finding solutions, although so far this is in quantum-inspired computing (see Chapter 8) more than in quantum annealers or gate-based quantum computers.

Quantum computing may offer your organization — or, if you don't step up, your competitors — new opportunities for competitive advantage very soon. Do you want to have a long learning curve in front of you when that moment arrives — if it hasn't already?

A clarifying question is to imagine a competing organization announcing a new initiative, new product, or new service based on quantum computing. This situation would create a lot of excitement for them — and a lot of pressure on you. The CEO or managing director is likely to ask their organization something like this: "How long will it take us to be competitive in quantum computing?" If you're in a responsible position, "I don't know" is not likely to be a satisfactory answer.

It's not that hard for an organization to create a part-time project team and take a serious look at what quantum computing has to offer at this point, as described in Chapter 8. We urge you to undertake such an effort before you find yourself and your organization missing out on opportunities — or, worse, left as roadkill on the quantum computing superhighway.

That's part of the reason we've been haranguing you in these pages about making an initial commitment of time and energy to quantum computing — the up front investment you'll need to make is not very large. In many cases, programs are available to help subsidize the upfront investment — such as the free hardware credits we've mentioned, and early access programs offered via AWS, Azure,

Strangeworks, and others — to help you and your team get started. If the field heats up, it may become harder and more expensive to get started.

The subtle advantage that the cloud gives you is in terms of supply. In industry terms, quantum computing is part of the high-performance computing (HPC) landscape. HPC used to be mainly about supercomputers: powerful standalone devices that cost millions of dollars. Making these systems required a major investment. An ecosystem developed in which serious vendors could count on a few initial orders from governments and research labs to help them get new and improved technologies off the ground, but this made for slow progress.

With the cloud, however, HPC is expanding quickly. Cloud vendors can make new and even exotic devices available cheaply, with immediate access to a large base of potential customers. The customers can "try before they (seriously) buy."

In Chapter 8, we described the emergence of quantum-inspired algorithms running on classical computers. This field has been able to grow quickly, and to offer a very wide range of solutions very early in its existence, because of flexible arrangements between customers and vendors that are possible only because of the cloud.

One company taking advantage of this situation is Nvidia, the chipmaker who used to be mainly in the graphics business. Today, their graphics-processing devices (GPUs) and related devices are in huge demand for AI and in growing demand for quantum-inspired computing. You may end up beginning your quantum computing journey on a simulator running on Nvidia GPUs.

Because a lot of complexity has sprung up fast, the minor downside for you, as a potential customer, is that not all the new solutions will last. But if you can identify your business challenges and opportunities, educate yourself about the broad trends in the industry — starting, perhaps, with a book like (ahem) this one — and assess new offerings efficiently, you can develop a potentially lasting competitive advantage as a smart user of this new technology.

Sharing a quiet word about companies

While we're on the topic of how quantum computing companies get born, we should also mention that they can die too.

Quantum computing is still in a boom-and-bust mode. Over the last few decades, as VC-funded AI companies have sprung up, we've seen several AI winters. An *AI winter* is when interest wanes, funding dries up, and several companies give up the ghost — or, perhaps worse, go into zombie mode. (Think of a mashup of the movie *Night of the Living Dead* and the TV series *Silicon Valley*.) In zombie mode, the lights are on, but there aren't many people home.

Quantum computing has had its quantum winters too. In fact, at a conference your authors attended in 2023, avoiding another quantum winter was one of the main topics of conversation — it even made the official agenda.

Check into the health of companies as part of your due diligence in assessing different modalities and products. Ask tough questions about revenues, customer wins, and business prospects.

The cloud is only somewhat helpful. It's easy for customers and companies to undertake initial efforts on a small budget. But customers don't have to commit to a company the same way they used to. In the past, when a supercomputer company gave you a customer list, those customers would each have spent at least several million dollars for at least one machine. Today, initial access is cheap and easy. So the customer list that a vendor shows you may be shallow in terms of total spending. (Right now, the biggest projects being awarded to quantum computing vendors are government contracts.)

For one example, a major quantum computing company sells quantum computing systems for more than $10M a pop. In recent years, their annual revenues have rarely reached — wait for it — $10M. So you can quickly assess that, no matter how strong the customer list may look, and in spite of impressive press releases, not a lot of those $10M-plus systems are heading out of the warehouse.

In fact, the warehouse probably doesn't exist. Instead, probably only a few of each newly released system are in existence, most or all in a vendor-owned facility, with fractional access to many customers made available through the cloud. This situation is good for customers who want to kick a lot of tires before committing but tough on a vendor's revenue numbers.

So ask those tough questions before you engage with a vendor. Whether the answers are reassuring or worrisome, have a backup plan in case a given supplier runs into headwinds.

3

Getting Entangled with Quantum Computing

IN THIS PART . . .

Program a quantum computer using steps described in this chapter, including free access to an IBM-powered introduction to quantum computer programming.

Find out what application areas for quantum computing show promise as the devices get more powerful, now and in the years ahead.

Study quantum computing algorithms that solve specific types of programming and real-world problems, each of which apply across multiple application areas.

Use the cloud to gain access to quantum computers and related resources through cloud providers such as AWS and Microsoft Azure and specialized providers such as Strangeworks.

Dig into educational resources that will provide you with everything from a gentle introduction at a high-school level to instruction in machine learning on quantum computers.

» Thinking like a developer

» Getting your quantum on

» Breaking down a quantum algorithm

Chapter 12

Programming a Quantum Computer

Programming a quantum computer is very different from programming a classical computer. If you're an experienced developer, you'll have some things to unlearn. If you're new to development, you'll have a steep learning curve but the advantage of a fresh start.

We recommend that you try quantum computer programming if you're interested in the field as a whole, even if you're not a dev and don't intend to become one. Why? Because quantum computing is outstanding at some things and not much use for others. There's no better way to develop a feel for what quantum computers can and can't do at their current state of development — and what they might be capable of in the future — than to learn at least some basic quantum computer programming skills today.

And if you are a dev, learning how to program a quantum computer is vital if you want to make quantum computers do much. You have to not only learn the basics but also gain some mastery of quantum computing programming to wring performance out of today's machines; it's a NISQy business. (Apologies to Tom Cruise, Rebecca DeMornay, and others involved in the 1983 movie *Risky Business*, and to you, our readers.)

REMEMBER

The NISQy reference is a bad pun on noisy intermediate-scale quantum (NISQ), the description we shared in Chapter 1 for today's ~~buggy~~ noisy and underpowered machines.

TECHNICAL STUFF

The effort to get quantum computers working productively is teaching developers and others a lot about current classical computers. We bet you'll develop a fresh perspective on your daily work programming classical computers by giving quantum computer programming a go.

At the end of this chapter, we've arranged to give you a gentle introduction to hands-on quantum computer programming, as inexpensively as possible. That is to say, free! (At least, free given that you've already paid for the book.) That way you'll have some momentum into the quantum computing programming courses that are out there, some of which we describe in Chapter 16.

WHY IS THIS SO DARN HARD?

Part of the reason why quantum computer programming is hard to learn is that it operates much closer to the hardware — spooky, as Albert Einstein said, as that hardware is — than does today's classical computer programming.

Programming for classical computers has progressed to such a high level that few programmers understand, or even know about, the classical logic gates that their programming languages resolve to. Many developers today don't even know what assembly language — the programming language that speaks directly to a CPU — is. They also don't know about the mathematical abstractions and physical implementations enabling their code to operate.

Back in the day — the 1940s and 1950s — to program then-new classical computers, you had to come correct, with a full understanding of the logical abstractions you were using and the electronic circuitry that executed your code. No longer.

As quantum computing progresses toward liftoff, we are again at a point where you need to understand the mathematical abstractions — the logic gates — and their implementation in circuitry (in this case, qubits) that enable your programs to operate. Did we mention that learning quantum computer programming may indirectly improve your classical computing programming skills as well?

Figuring Out What We Are Doing

tl;dr optimizing

REMEMBER

tl;dr is internet jargon for "too long; didn't read." It's a pithy summary, prepended to lengthy material that you may or may not then choose to go ahead and read.

One of the authors (Smith) was an early employee in a 1990s startup, now lost to memory, called Taligent. The new company adopted a motto: We Only Do What Only We Can Do, abbreviated WODWOWCD. And yes, they had T-shirts printed, and did several years of work with WODWOWCD as their North Star.

Unfortunately, the motto, the T-shirt, and a CD-ROM — remember those? — with all our code and documentation, which sold for $100, were the only valuable outputs from Taligent. (We need an Ozymandias-type poem for tech startups. "Look on our code, ye mighty, and despair. . . .")

But the motto is pretty cool! It tells us, in a complex situation, to change not all the things we can change but only the things that we are *uniquely positioned* to change. And quantum computer programming is kind of like that. In these early days, you can pick up these new tools and attack challenges in a way that no one may have ever done before.

As quantum computing matures, early experience is likely to give you an advantage when a situation has an immense number of possible answers and you need to do one or more of the following:

- Find a better answer (optimization), the focus of the examples in this chapter
- Find the best answer (calculation)
- Understand and change the behavior of matter when it's acting according to quantum mechanical principles (simulation)

TECHNICAL STUFF

We show you here how to use optimization to generate answers to problems. Within optimization, you're currently likely to get better results using quantum-inspired computing (Chapter 8) or quantum annealing (Chapter 9) than logic-gate quantum computing (Chapters 10 and 11). But programming a gate-based quantum computer is one heck of a learning experience now, and may well prove valuable to you in the future.

What we ask you to keep in mind is that today's quantum computers only have enough power to start you up the slope of the Mount Everest that a given problem represents. There are few problems, and no truly challenging ones, where we have enough quantum computing power at this point to reach the peak — that is, to achieve quantum advantage.

THE POWER OF QUANTUM COMPUTING

The power of quantum computing lies in massive parallelism. When we have fully capable quantum computers, they'll operate exponentially faster than classical computers for the problems quantum computers are good at.

A quantum computer with, for instance, 256 error-corrected qubits — which may require 100 or 1,000 times that many physical qubits, far more than we have today — will be able to find the right answer to problems with 2^{256} possible answers. Adding just one qubit will double the number of possible answers to 2^{257}, and so on as we add qubits.

So let's say it takes a classical computer 1 second to come up with a single answer to a problem that has 2^{256} possible answers. It will then take another second to come up with the next answer. And so on; you have to run the problem through the classical computer 2**256 times to generate all the possible answers.

Unfortunately for the user, 2^{256} is a very large number, and it will take the classical computer longer than the expected age of our universe to calculate all the possible answers to that problem — answers that would be of much greater value to the user if they were still alive and kicking.

But let's say that it also took a quantum computer with 256 error-corrected qubits 1 second to solve the same problem. However, the quantum computer will not come back with *one and only one of the possible answers* to the problem, as a classical computer would, and make the user repeat the process 2^{256} times. A mature quantum computer will come back with *the best possible answer*, out of 2^{256} options, to the problem, on the first try. One second should lie well within the user's life expectancy!

If this improved quantum computer of the future was somewhat buggy, you could still get the right answer. You would need to run the problem quite a few times and wait for the right answer to appear repeatedly. After you got a statistically significant number of hits on the same answer, you could declare victory — after some hassle, but still well in advance of the heat death of the universe. If you spend much time around today's quantum computers, you'll see this technique in use on smaller problems.

THE RACE FOR HPC SUPERIORITY

As fascinating as quantum computing is, it's also part of other, larger pictures. One is Quantum Technology 2.0, where quantum computing sits cheerfully alongside quantum communications and quantum sensing. Another is the bigger picture around high-performance computing (HPC), as mentioned in Chapter 8.

If data is (still) the new oil, compute is the new electricity. Solid progress has been made in using HPC for solving humanity's problems and in gaining a competitive advantage for organizations. To HPC types, quantum computing is just another cool tool in Batman's utility belt.

If you get seriously involved in quantum computing, you're likely to learn a lot about the broader HPC picture. That's because various kinds of quantum-inspired computing running on HPCs often rival quantum algorithms on various metrics; speed, cost, and energy consumption lead the key performance indicators. You may need to try a variety of approaches to see whether quantum computing really is the way to go for your problem.

We expect many people who are initially attracted to quantum computing to become all-around HPC mavens, advancing computing on all platforms.

REMEMBER

Quantum advantage means getting better — more accurate, faster, or less expensive — results from a quantum computer than from a classical computer. We don't have quantum advantage yet. For instance, an early demonstration of Shor's algorithm on a quantum computer (see Chapter 6) was able to find the prime factors of . . . 15. Sorry, but that's the world we're living in, and if you get involved in quantum computing, it will be your job to help make it suck less.

Figuring Out How to Do It

tl;dr algorithms and gates

In classical computing, the developer tells the computer to take steps toward a desired solution. In quantum computing, we describe the problem to the computer and use programming to create an environment in which the desired solution is returned to us.

As with other quantum mechanical phenomena, the reason that quantum computing returns correct results — even for trivial problems like factoring 15 — doesn't make sense to our limited primate brains. But when in doubt, as Richard Feynman once said, "Shut up and calculate."

Quantum mechanics returns incredibly useful results to ridiculously hard problems with amazing precision, and has been doing so for more than a century. (Specifically, at least since Albert Einstein's 1905 paper on black body radiation. Thank you, Uncle Albert.) Quantum computing should at some point do the same, whether or not we ever really understand why.

You have two key tools in attacking problems with quantum computing. The first is quantum algorithms. Quantum algorithms find the part of a problem that takes incredibly long to calculate and use quantum computing's parallelism to return the answer to that part quickly. (As quantum computers get more powerful, the answer will be returned *very* quickly.)

We describe several key algorithms and how they work their magic — and what they do *is* kind of magical — in Chapter 14. For now, just think of these algorithms as scalpels.

You can't solve every problem with a scalpel, but when needed, scalpels are useful indeed. Much of your work as a quantum computing programmer will be to identify problems that can be attacked with one of the algorithms — scalpels — that you have available, and then to wield your chosen algorithm effectively.

The other tool you have are the actual steps (okay, steplike things) in a quantum computing program, which are called gates. These are logic gates, analogous to those in classical computing (which few of us deal with directly) but different, in interesting quantum ways.

The most consequential difference is that quantum computing logic gates are reversible. That is, you always have an undo button for each step in a quantum computing program. It takes a new and different mindset to solve problems using reversible steps, and you'll develop that mindset as you do quantum computer programming.

Using reversible steps to accomplish computational work dances along the edge of falling into the deep, dark abyss just beyond the Second Law of Thermodynamics, which states that overall disorder inexorably increases. (Which also means that your existence, and each of ours, is just a small speed bump on the way to ultimate chaos. This also happens to be the theme of the better Marvel superhero movies, but whatever.) When you use reversible steps in your quantum computing programs, you don't decrease overall disorder, which is theoretically impossible, but you don't increase it either. Pretty cool, huh?

Part of the reason for doing quantum computer programming is so you can develop this new mindset. As we mention at the beginning of the chapter, using this new mindset is helpful not only in programming a quantum computer but also in identifying problems that are ripe for quantum computing to solve. As the power of quantum computers increases in the years ahead, they will be able to solve an ever-widening range of problems.

Understanding what quantum computing logic gates do requires some reflection on the deeper meaning of imaginary numbers, an understanding of matrix math, and informed use of a Bloch sphere. We aren't going to go through all that carefully in this book, but you should expect to learn it in a good introductory quantum computing programming course. Even so, we can't resist showing you a bit of what it's all about.

Many of us were introduced to matrix math in high school or university coursework but then lost whatever skill we had developed due to disuse. If you want to take up quantum computer programming as a regular practice, take the time to learn, or re-learn, matrix math; it's comprehensible to most mere mortals and crucial to unlocking the power of quantum computing.

Figure 12-1 shows a representation of a Bloch sphere. Bloch spheres are useful for visualizing what quantum computing logic gates do. They're also extremely cool, and you can almost certainly get your head around them if you spend some time with them.

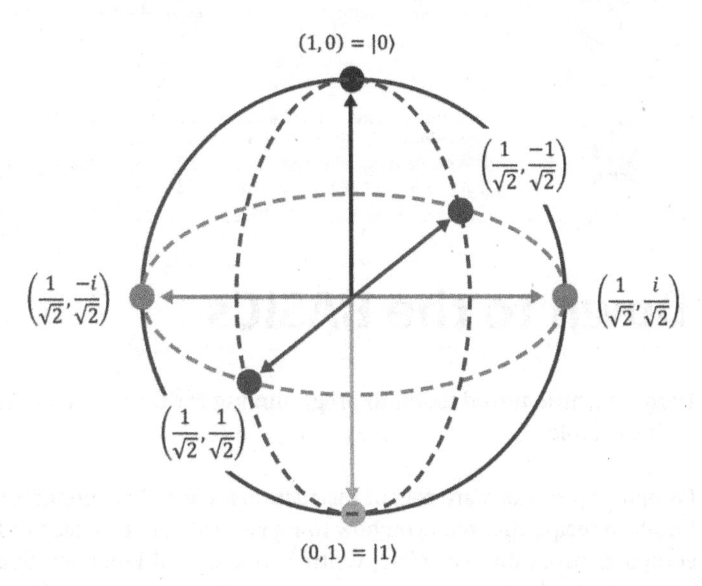

FIGURE 12-1: A Bloch sphere.

TECHNICAL STUFF

Geomancy is communicating with the supernatural by interpreting patterns or markings on the ground, whether these occur naturally or are created by tossing out rocks, sand, soil, and the like. Quantum computing's use of the Bloch sphere is not that different, except for being in 3D and incorporating imaginary numbers. (And only in a *Dummies* book about quantum computing can a comment about geomancy be labeled as technical stuff.)

REMEMBER

Taking us back to high school geometry here, math for a Bloch sphere uses radians. A circle is 2π radians around, so $1/2\pi$ radians is a 90-degree shift, π radians is a 180-degree flip, and 2π radians is a flip all the way around the circle and back to where you started.

Figure 12-2 shows a few quantum computing logic gates, which you can visualize by imagining what they do when mapped to a Bloch sphere. You'll learn about a more complete set of logic gates as you learn quantum computer programming.

Logic Gate	Functionality
X	X-gate. Rotates π radians around the X axis. Equivalent to a bit flip in classical computing.
Y	Z-gate. Rotates π radians around the Z gate. This is referred to as a phase flip and does not have an equivalent in classical computing.
H	Hadamard gate. Puts the qubit into a state of superposition, such that the orientation of the Bloch sphere becomes indeterminate. Reading the qubit is equally likely to yield a 0 or a 1. This also does not have an equivalent in classical computing, at least not if you're sober.

FIGURE 12-2: Selected quantum computing logic gates.

Getting Down to the BASICs

Here's a gentle introduction to programming for those of you who are unfamiliar with the topic.

Computer programs are sets of instructions that tell a computer what to do. It's a bit like a recipe that tells you how to make a cake; but instead of flour and eggs, a computer program uses code, written in a special language that computers can understand.

A program starts with the input, which is data that the computer needs to know to do its job. Input might include numbers, words, or pictures. Next, the program uses logic to process the input and decide what to do with it. This step might involve calculating, checking if certain conditions are met, or making decisions based on the input.

Once the program has processed the input, it produces output, the result from the program's calculations. The output might include numbers, words, pictures, or instructions to another program, depending on what the program was intended to do.

Quantum computers currently do only the processing part. The input will be brought to the quantum computer by a classical computer, and the output will be sent from the quantum computer to a classical computer.

Programs can do all sorts of things, from simple calculations to complex tasks like creating a video game experience for game players or analyzing data. Programs can be written in different programming languages and can run on different types of computers, from phones to desktops to supercomputers to, well, quantum computer simulators or quantum computers.

Learning to write computer programs is a fun and useful skill that can open up all sorts of opportunities for you.

Writing the Requirements for a Quantum Program

The bits used in classical computing have defined values, either 0 or 1, and each bit is fully independent of every other bit. A qubit, used in quantum computing, can be in several states at once (superposition), and the state of each qubit can interact seamlessly with one or more other qubits (entanglement), making qubits far more powerful than bits. Quantum computing programs can also home in on the correct answer from a wide number of possible options (tunneling).

Despite the differences between classical and quantum computer programs, a quantum program is simply a set of instructions for a quantum computer. In a quantum program, you tell the quantum computer what to do by creating a series of instructions. These instructions are called quantum gates.

A gate in quantum computing serves roughly the same purpose as a step in computer programming. (A quantum computer programming gate is the same as a classical computer programming step, except the gate is reversible and accesses the multiverse, which means it's very different.) Gates are fundamental operations closer to the processor-specific instructions in assembly language than to the high-level instructions in most classical computing programming languages.

To make quantum computer programming (much) easier, existing classical computer programming languages, such as Python, have been adapted to work with quantum computer programs. However, most Python programmers don't understand what their programs do at the machine level when run on a classical computer, and they sure as shell scripting don't understand exactly how their programs interact with qubits. Be prepared to dig deep if you don't want to be some kind of McProgrammer of quantum computers — someone who can get a program to run without really understanding what it does.

You can think of quantum gates as dials and switches that change the state of a set of qubits. Developers can combine these in different, seemingly simple ways to create complex operations. When a quantum program is executed (run), the quantum computer sends gates to manipulate the qubits and perform the task you asked for.

Gates change the state of qubits in subtle ways. But when the program is complete, the results are read out as either a 0 or 1 for each qubit, similar to the results we see when programming classical computers.

The best way to understand what this all means is to write and run some quantum computing programs. The first thing you need to get started is a proper development environment. You can choose from a number of environments; if you don't have one, look into Jupyter Notebooks, Microsoft Visual Studio Code, or XCode.

You'll also need support beyond this book. StackOverflow, GitHub Copilot, and even ChatGPT are free or inexpensive resources that can help you along the way. Chapter 16 gives you some pointers.

Components of a quantum programming environment

To create your first quantum program, you need a few things, some of which are tangible, such as programming environments and languages, and some of which are intangible, such as knowledge of quantum gates and algorithms:

- **A quantum programming framework:** The programming framework consists of a set of tools and libraries to create and run quantum programs. Some examples of quantum programming frameworks are IBM's Qiskit, Rigetti's Forest, Google's Cirq, Xanadu's Pennylane, and Microsoft's Quantum Development Kit. Qiskit is currently the most popular.
- **A quantum simulator or quantum computer:** To run your quantum program, you need access to a quantum simulator or, if you're lucky, a real-life quantum computer. Quantum simulators are software tools that can simulate the behavior of a quantum computer, while real quantum computers are physical machines that use qubits to perform computations. The other main difference is cost. You can quickly gain access to simulators; quantum computers can be cost-prohibitive. (Free credits are often available to cover the cost of your initial efforts on a real quantum computer, but using one can get expensive after that.)
- **A knowledge of quantum gates:** Quantum gates are the basic building blocks of quantum programs. To create a quantum program, you must know how to use various quantum gates, such as the Hadamard gate, CNOT gate, and phase gate, to manipulate qubits to perform quantum operations.
- **An initial understanding of quantum algorithms:** Quantum algorithms are specialized methods for solving problems using quantum computers (see Chapter 14). Depending on the task you want your quantum program to perform, you may need to know about one or more specific quantum algorithms, such as Shor's algorithm for factoring large numbers or Grover's algorithm for searching unsorted databases.
- **Basic programming skills:** You need programming skills to write your quantum program. Most quantum frameworks use programming languages such as Python or Q# to create and run quantum programs. Therefore, you must know how to use these languages to develop software. Or you must be willing to work hard to learn basic programming skills as you write your initial quantum computing programs; many of the great innovators in tech learned as they went along.

Components of a quantum program

Now that you know what your development environment requires, let's review the necessary components to run your new program. Quantum programs consist of several items that work together to perform quantum computations. The critical components of a quantum computer program include the following:

- **Qubits:** Qubits are the fundamental building blocks of quantum computers. They're similar to the bits used in classical computers, but they can exist in multiple states simultaneously and can be entangled with each other, allowing quantum computers to perform calculations in parallel. In a quantum program, you use qubits to process information.

- **Quantum gates:** Quantum gates are the operations that can be performed on qubits to manipulate their state. They're similar to logic gates in classical computers, but they take advantage of the unique properties of qubits, such as superposition and entanglement. Standard quantum gates include the Hadamard, Pauli-X, and CNOT gates.
- **Quantum circuits:** A quantum circuit, as shown in Figure 12-3, is a sequence of quantum gates applied to one or more qubits to perform a computation. In a quantum program, you typically create a quantum circuit by specifying the sequence of gates you want to apply to your qubits.
- **Initialization:** Quantum gates need to be initialized, put into a state of superposition, and entangled appropriately with other qubits before a program can be run. This process is called initialization.
- **Measurements:** Taking a measurement is how you extract information from a qubit. When you measure a qubit, it collapses — the technical term is *decoheres* — into a 0 or 1, which you can then read as a classical bit.
- **Error mitigation:** Because quantum computers are sensitive to noise and other sources of error, many quantum programs include error correction routines to improve the accuracy of the results. These routines typically involve adding extra qubits and gates to the program to detect and correct errors. However, they are not 100 percent effective, or even close, unlike error correction on classical computers.

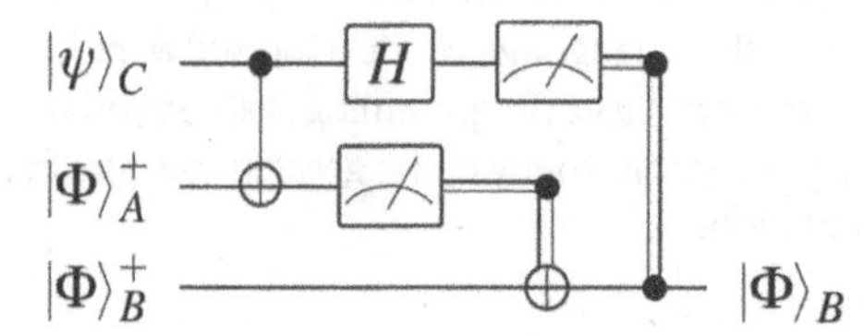

FIGURE 12-3: An example of a quantum circuit.

Quantum programs are much like classical programs, except they take advantage of the unique properties of quantum systems to perform computations in new and exciting ways. When you develop a quantum program, you enlist qubits, gates, circuits, measurements, and error mitigation techniques to use quantum mechanics to your advantage.

Gates are like special tools that can change the position of the qubits. So, for example, a gate might rotate a qubit from on to off, or from off to "on and off at the same time." When you connect the qubits and gates in a circuit, you can perform calculations and solve problems that are very difficult or impossible to solve with regular computers. For example, you might use a quantum circuit to factor a large number or to search through a large database of information.

Thinking Like a Developer

This section is for beginners, but current devs might be interested in this section as a quantum-flavored review.

If you want to write a quantum program, you'll need to gain a firm grasp of software development: what it is, how it works, and what to watch out for as you start your journey.

Software developers usually work on a team with other developers, designers, and project managers. They use their skills to help create software that meets the needs of their clients or users. They might work on different parts of a program, such as the front end (what people see on their screens) or the back end (the part that runs on servers).

Since you're programming a quantum computer and writing simple programs, you'll focus mainly on setting up your development environment and getting your initial program to work. But before you get to that, let's discuss what makes an awesome software developer.

Tips for becoming an awesome software developer

To become a software developer, people usually need to have some sort of training or education in computer science, such as going to college or a university, taking online courses, or attending a coding camp. They might also need to practice and learn on their own by writing their own programs and experimenting with different tools and languages.

However, many software developers started with little formal training (like both of the authors). Either way, you can become a developer regardless of your prior experience. Here are a few things to keep in mind as you start (or continue) your journey:

- **Write clean, well-organized code.** Your code should be easy to read and understand. Use consistent formatting, meaningful variable names, and explicit comments to make your code easier to follow.
- **Test your code thoroughly.** Before deploying code, test it to catch any bugs or errors. Use manual testing methods and automated testing tools to ensure that your code works correctly.
- **Use version control.** Version control is a system for tracking changes to your code over time. It allows you to keep a history of your code, collaborate with other developers, and go back to previous versions if needed. Git is a popular version control system used by many developers.
- **Document your code.** Documenting your code with comments and other documentation helps other developers understand your code and how to use it. Documentation also enables you to remember why you made certain decisions or wrote certain pieces of code in the way that you did.
- **Keep learning.** Technology constantly evolves, so it's essential to keep learning and staying up-to-date with the latest trends and best practices in software development. Attend conferences, read blogs and articles, and participate in online communities to stay informed and connected with other developers.
- **Collaborate.** Collaborating with other developers is a great way to learn new skills, get feedback on your code, and share knowledge. Joining open-source projects or contributing to online forums and communities can help you connect with other developers and build your skills and experience.
- **Use first principles during development.** In software development, first principles refer to the fundamental building blocks and concepts that underlie the development process. By focusing on the building blocks and underlying principles, developers can create more resilient, adaptable, and innovative software that can ultimately drive progress in the field.

LIKE A VERSION

Madonna did not sing, "Like a version, debugged for the very first time," but she might have done so if she were a developer. In software, version control is how developers track changes to their code over time. We recommend that you save versions of your code early and often.

Version control allows developers to keep a record of each change made to their code, including who made the change, when it was made, and what was changed. This makes it easier to collaborate with other developers, track the history of the code, and revert to previous versions if necessary.

The most common version control system used by developers, and the one the authors strongly suggest is Git, which was created by Linus Torvalds back in 2005.

One of the key benefits of version control is that it allows developers to work on the same codebase at the same time without overwriting each other's changes. Rationalizing the changes is done through merging, which is the process of combining different versions of the code.

When two developers make changes to the same piece of code, Git automatically attempts to merge the changes. If conflicts exist between the changes, Git will highlight them so the developers can resolve them together.

Another benefit of version control is that it provides a history of all changes made to the code. This record can be useful for debugging issues, tracking down bugs, and understanding why certain changes were made. But any way you cut it, trust us, you'll save yourself a lot of future heartaches if you version early and often.

Setting Up a Development Environment

One of the most popular languages for programming quantum computers is Python, so you'll typically create your example quantum programs using this language.

To get started with Python, install it using the app download available at `www.python.org/downloads/` and follow their instructions. Once Python is installed, you'll need to set up an account at Strangeworks, a quantum computing company that provides access to most of the available quantum hardware and software available today. (One of the authors, whurley, founded and is CEO of Strangeworks.)

The Strangeworks quantum computing platform ties all the different pieces of the quantum computing ecosystem together so you can get up and running as easily as possible.

Even if you consider yourself a novice coder or have no experience, we want to encourage you to take a leap of faith and go through these steps. Start by creating an account at `https://strangeworks.com/`. From this account, you can browse your quantum jobs, view available quantum computers, activate hardware, and retrieve your API key.

Next, you need to choose an integrated development environment (IDE). An IDE is like a special toolbox that helps you build your program. Just like you use different tools for different jobs, you use an IDE to write, test, and run your code.

If you're a new developer without experience, you might feel overwhelmed by all the different IDEs. The first thing to do is ask yourself what you want to create. For example, if you want to create a website, you would choose an IDE specifically designed for web development. Or, if you want to create a game, you might choose an IDE with built-in game development tools.

Once you have an idea of what you want to create, you can start exploring available IDEs. Some popular ones include Visual Studio Code, PyCharm, and Jupyter Notebooks. The authors encourage you to consider this last option because it includes automation capabilities.

It's also good to read reviews and watch tutorials to see what other developers recommend. And remember, as you learn and grow as a developer, you might prefer one IDE over another at different times. Don't be afraid to try out different IDEs until you find one that feels comfortable and helps you create the things you want!

When you have installed Python, created a Strangeworks account, and selected an IDE, you'll be ready to program a quantum computer.

Finding Where to Get Your Quantum On

You need more than a laptop and a programming language to program on a quantum computer. That's why the friendly folks at Strangeworks have created an excellent resource for this book. You can find it online at `www.dummies.com/go/quantumcomputingfd`.

NEW TO QUANTUM? HAVE A QISKIT

If you're new to quantum computing development, we recommend starting with IBM's Qiskit, one of the most popular open-source frameworks for quantum computing. Qiskit helps you create programs for quantum computers using Python.

With Qiskit, you can create quantum circuits that use qubits to solve complex problems much faster than regular computers. Qiskit also includes a simulator that enables you to test your quantum circuits on a classical computer, even if you don't have access to a real quantum computer.

But the best part is that Qiskit is available for anyone to use for free. Qiskit is an exciting new tool that empowers you to explore the amazing world of quantum computing.

Getting started with the Strangeworks Python SDK

The Strangeworks Python SDK handles authentication, fetches jobs and backends, transfers files, and calls product endpoints.

Installation

To get started, make sure you have Python 3.9 or later. You can easily install the Strangeworks SDK from PyPI with the pip command:

```
pip install strangeworks
```

Authentication

To access the Strangeworks API, you'll need your API key, which you can find in the Strangeworks portal on the home page. Once you have it, simply authenticate like so:

```
import strangeworks

strangeworks.authenticate(
    api_key="your-api-key"
)
```

If you belong to multiple workspaces, please be aware your API key for each is different. If you do have multiple workspaces, you may be getting a little ahead of us at this point, but the Strangeworks documentation is there to help you sort things out if you run into trouble.

While you can use the Strangeworks Python SDK package directly, it's most commonly used as a dependency of task-specific packages such as Qiskit, Rigetti pyQuil, and Amazon Braket. You can find documentation for each of these packages on their product pages in the Strangeworks ecosystem.

Let's optimize something

The Strangeworks Python SDK allows users to submit quadratic unconstrained binary optimization (QUBO) problems using quantum computing through a REST API.

QUBOs are readily handled by quantum simulators and by quantum-inspired algorithms on classical computers (see Chapter 8), as well as by quantum annealers such as D-Wave's machines (see Chapter 9). For logic-gate quantum computers, optimization problems are not formulated as a QUBO but typically as a QAOA (quantum approximate optimization algorithm) or another type of input.

If you seek to put a QUBO workload into production, we recommend that you first conduct benchmarking to compare its price and performance against those of classical computing, as well as on different types of quantum computing platforms.

The Strangeworks Quantum Optimization Service lets you use QUBO problems as input, run the optimization problem using various backend solvers, and then receive the results as a standardized output file. The service makes it easy to run an optimization problem across different backends without needing to use different input/output schemas or multiple API keys.

To use the Strangeworks Optimization Service, make sure you have Python 3.9 or later and are familiar with setting up and using virtual environments.

Install the needed packages using pip:

```
> pip install strangeworks
> pip install dimod
```

Next, you import the package into Python:

```
import strangeworks
from dimod import BinaryQuadraticModel, SampleSet
import json
```

In the Strangeworks portal, activate Strangeworks Optimization Service to create a resource.

Now authenticate and configure the Python SDK, replacing *your-API-key* with your key from the Portal home page:

```
strangeworks.authenticate('your-API-key')
```

To make sure everything goes smoothly with your first quantum programming experience, please make sure you have done the following:

- Set up your environment and install Strangeworks.
- Activate the Strangeworks Optimization Service to create a resource.
- Replace *your-API-key* with your key from the Portal home page.

Now you can use the upcoming examples to run your first quantum program.

Let's kick things off by getting the resources we need:

```
resource = strangeworks.get_resource_for_product("optimization")
```

Now, you're ready to put all this together by submitting your QUBO problem:

```
path = "qubo"

linear = {1: -2, 2: -2, 3: -3, 4: -3, 5: -2}
quadratic = {(1, 2): 2, (1, 3): 2, (2, 4): 2, (3, 4): 2, (3, 5):
  2, (4, 5): 2}
bqm = BinaryQuadraticModel(linear, quadratic, "BINARY")

qubo_job = {
    "bqm": json.dumps(bqm.to_serializable()),
    "var_type": "BINARY",
    "lagrange_multiplier": 1.0,
    "solver": {
```

```
        "solver": "azure.Tabu",
        "solver_options": {}
    }
}

result = strangeworks.execute(resource, payload=qubo_job,
   endpoint=path)
sample_set = Sample.Set.from_serializable(json.
   loads(result["samples"]))

# Print solutions
 for s in sample_set:
     print(s)
```

If everything went well, the results will appear onscreen. And you will have successfully used the Strangeworks Optimization Service to solve a QUBO problem! How exciting is that?

If something goes wrong, use the link in the platform to join the community in Slack, which is full of people, including members of the Strangeworks team, to help you get things up and running. Community is one of the most important aspects of exploring a new technology, and you shouldn't feel ashamed to ask the internet for help. Everyone has to start somewhere; you're just starting in a much cooler place than most.

Singing QAOA-ooooo

Sometimes we run into problems that seem insurmountable. They're computationally complex and require extensive computer resources. In these cases, it's often helpful to break the problem down into smaller, more manageable parts. Think of it like a jigsaw — if you have a really hard puzzle to solve, you might break it down into smaller sections and solve each section separately. Then, you put all the sections together to solve the whole puzzle.

A quantum approximate optimization algorithm (QAOA) is a special type of algorithm that solves really hard problems by using a logic-gate quantum computer. It's like a superhero that comes to the rescue when regular algorithms can't get the job done. A QAOA breaks down the problem into smaller pieces that are easier to solve, uses quantum computers to solve each of these pieces, and then puts the pieces back together into a resulting solution.

These algorithms are really cool because they have the potential to solve problems that regular algorithms can't, and to do it much faster than regular computers. This potential will be realized more and more fully as quantum computers become more powerful in the years ahead.

Enter the strangest QAOA you've ever seen. To get started on this QAOA adventure, you go back to the Strangeworks Ecosystem but this time do things a little bit differently. Strangeworks provides the strangeworks-qaoa package, which enables you to run, construct, and solve problems with the QAOA algorithm on multiple hardware providers.

The QAOA algorithm allows you to solve quadratic unconstrained binary optimization (QUBO) problems by making use of the power of quantum resources. It's a hybrid classical-quantum computing algorithm. The general idea is that we propose a quantum ansatz circuit with some variational parameters as the coefficients for some of the quantum gates. We then apply this circuit to our qubits, measure the output, and repeat many times to build up the required statistics.

We take these measurement results and compute the cost function with our defined problem. We then hand the algorithm over to a classical CPU, which optimizes the circuit parameters before passing back to the quantum side to remeasure the cost function, but now for the new parameterized quantum circuit. The algorithm then iterates back and forth between the application of the quantum circuit and the classical optimization of the circuit parameters before eventually converging on a solution with the lowest cost.

Getting started with QAOA

For this example, you utilize the QisKit runtime so that you can run jobs through IBM's free services. Python 3.9 or later must be installed on your machine, but if you followed the first exercise, you've already installed it.

Before you can run anything you need to make sure that you install the strangeworks-qaoa into your development environment. You can do this in four easy steps:

1. **From the left-hand navigation, choose the Strangeworks portal. Then sign in.**
2. **Navigate to the catalog, as shown in Figure 12-4.**

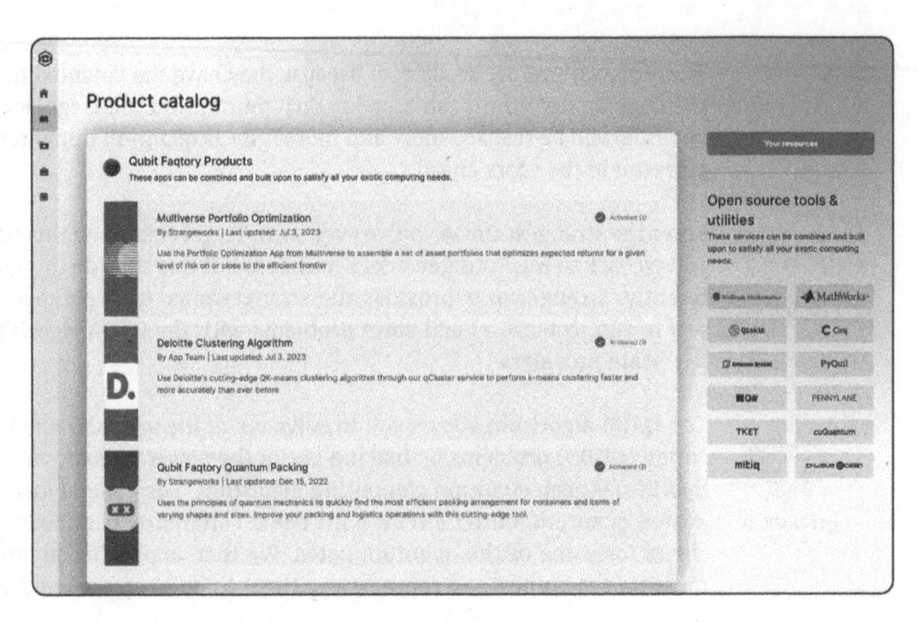

FIGURE 12-4: The Strangeworks product catalog.

3. **Using the icons on the left side of the screen, locate and click Strangeworks QAOA Service.**

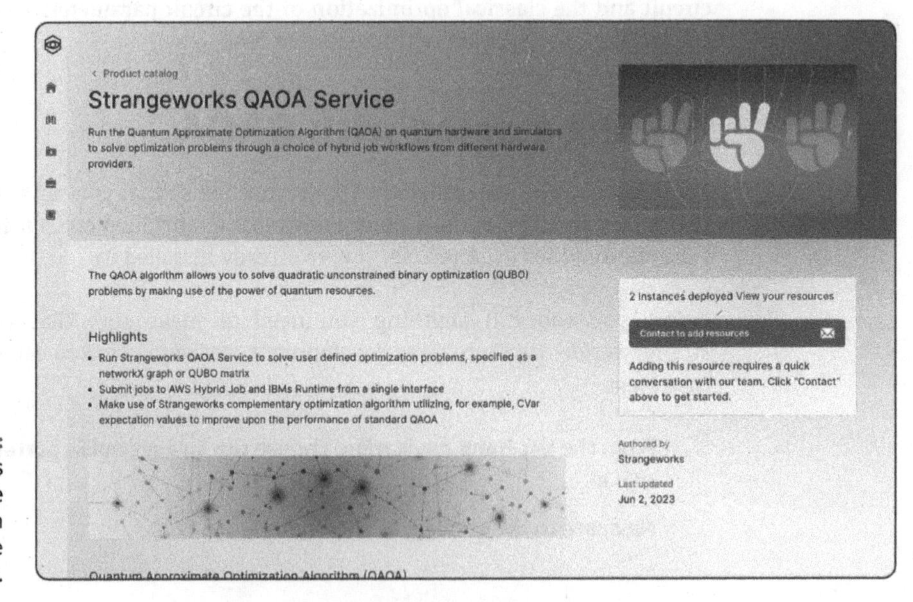

FIGURE 12-5: The Strangeworks QAOA service makes simulation more approachable.

The screen shown in Figure 12-5 appears.

4. **On the right of the screen, click the Contact to Add Resources button to get assistance in adding the service to your development environment.**

   The screen shown in Figure 12-6 appears.

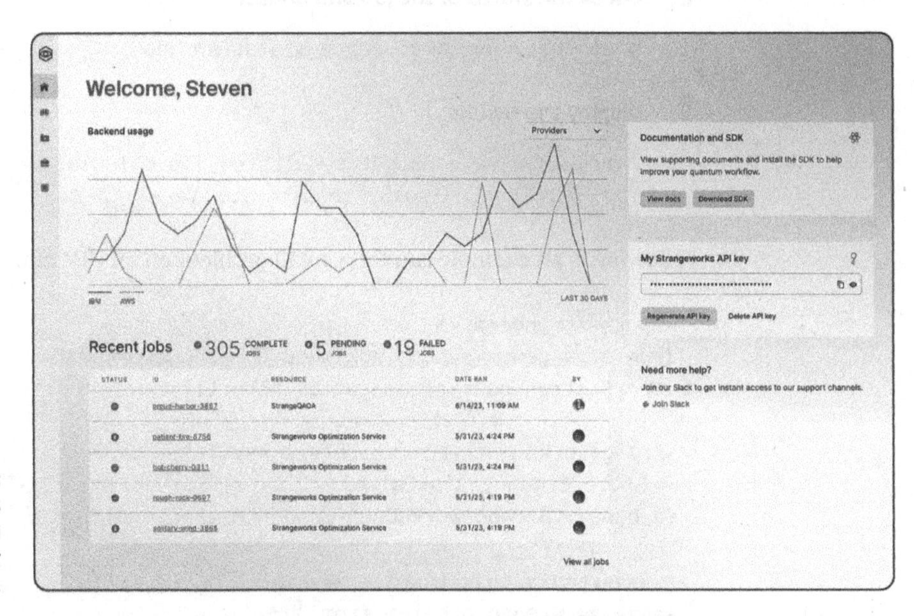

FIGURE 12-6: This screen is where you replace your API key.

Now please follow these steps to exercise the various capabilities of this service:

1. **Authenticate with the Strangeworks Python SDK using your API token:**

```
strangeworks.authentication(api_key='your-API-key')
```

2. **Get the QAOA resource using the QAOA SDK extension and your resource slug:**

```
sw_qaoa = StrangeworksQAOA(resource_slug=" ")
```

3. **List compatible backends for the QAOA product:**

```
sw_qaoa.backends()
```

4. **List all the user's QAOA jobs:**

```
jobs = sw_qaoa.job_list(update_status=True)
```

5. **Run the problem:**

```
sw_job = sw_qaoa.run(backend, problem, problem_params)
```

6. **Retrieve the job with the specific slug:**

```
sw_job = strangeworks.jobs(slug-slug)[0]
```

7. **Check the status of the job and update:**

```
status = sw_qaoa.update_status(sw_job)
```

8. **Display the results:**

```
result = sw.qaoa.get.results(sw_job,calculate_exact_
    sol=True, display_results=True)
```

Following is an example run for a small problem on an IBM simulator:

```
import strangeworks
from strangeworks_qaoa.sdk import StrangeworksQAOA
import strangeworks_qaoa.utils as utils

strangeworks.authentication(api_key=" ",
   store_credentials=False)
sq_qaoa = StrangeworksQAOA(resource_slug=" ")

####################################
######## Create problem from QUBO
nodes = 4
seed = 0
n = 3
problem = utils.get_nReg_MaxCut_QUBO(n, nodes, seed)

maxiter = 50
shotsin = 1000
theta0 = [1.0, 1.0]
p = 1
alpha = 0.1
optimizer = "COBYLA"
ansatz = "qaoa_strangeworks"

problem_params = {
    "nqubits": nodes,
    "maxister": maxiter,
    "shotsin": shotsin,
```

```
    "theta0": theta0,        #optional
    "p": p,                  #optional
    "alpha": alpha,          #optional
    "optimizer": optimizer, #optional
    "ansatz": ansatz,        #optional
    "ising": False,          #optional
    }

backend = "ibmq_qasm_simulator"
sw_job = sw_qaoa.run(backend, problem, problem_params)

result = sw.qaoa.get_results(sw_job,calulate_exact_sol=True,
   display_results=True)
```

Success! You're now ready to view your job in the Strangeworks portal, as shown in Figure 12-7, with the new job at the top of the list.

So, what exactly did you just do? Great question! Let's chat about that in the next section.

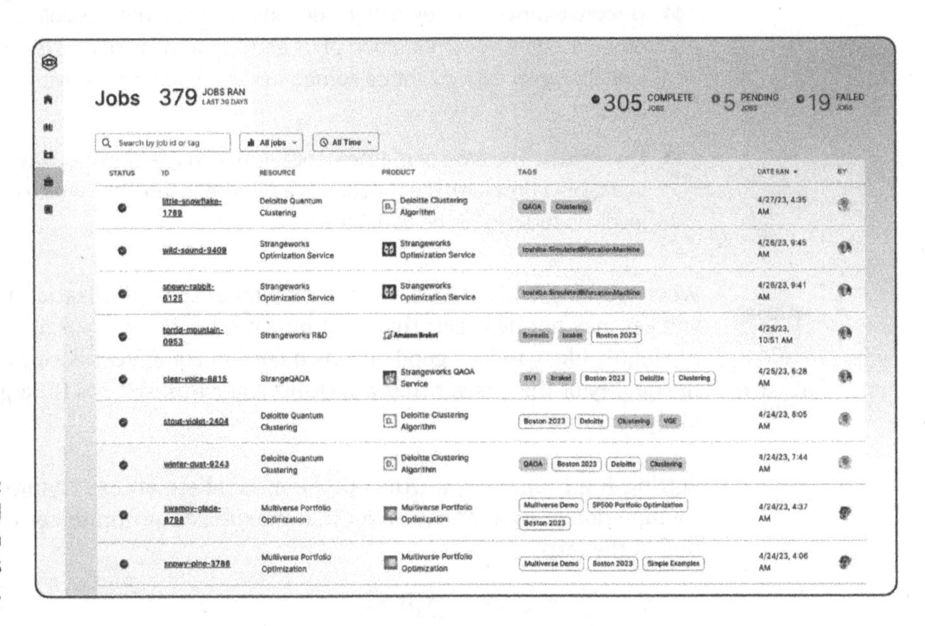

FIGURE 12-7: The completed job appears in the Strangeworks portal.

Breaking Down a Quantum Algorithm

Congratulations! You just ran a QAOA on an actual, real-life, quantum computer! How cool is that? But what exactly happened? Well, let's break it down.

So how does a QAOA work again? The QAOA algorithm solves a quadratic unconstrained binary optimization (QUBO) problem. As a reminder, QUBO problems are a special type of optimization problem. The main words here are *quadratic*, *unconstrained*, *binary*, and *optimization*. From the last to the first, they are described as follows:

- **Optimization:** Finding the best solution to a problem, such as the shortest path between a set of cities, minimizing the amount of resources used in a process, or maximizing profit in a business.
- **Binary:** When the variables in the problem can have only one of two values, often represented as 0 or 1. For example, in a problem where you're trying to decide whether or not to include a city in a travel itinerary, a city could be included (1) or not included (0).
- **Unconstrained:** No restrictions or constraints on the possible solutions. In other words, any combination of 0s and 1s for the variables is allowed. If there were constraints, it might be something like "you have to visit at least three cities."
- **Quadratic:** The type of mathematical function used to calculate the solution. In a QUBO problem, the function is a quadratic function, which means it involves squares of the variables.

When working with QUBOs, always remember the optimization part. The answer you get may not be exact; that is, it may not be the best possible answer, though it's highly likely to be a good one. As quantum computers become more powerful, the quality of the answer that the QUBO algorithm returns for a given problem is likely to improve.

Putting it all together, a QUBO problem is where you're trying to find the best combination of 0s and 1s (binary) that maximizes or minimizes a quadratic function of the form

$$C = \sum_{n,m>n} A_{n,m} x_n x_m + \sum_n B_n x_n,$$

where

$$x_n = \{0,1\},$$

does not have any restrictions or rules (unconstrained) about which combinations are allowed.

Now let's stick to the problem of finding the shortest possible route between two points, a problem often referred to the "traveling salesman problem." This problem is popular in the quantum computing literature because it's used in many real-world situations and because the number of possible solutions increases exponentially with the number of points that can be visited.

The QUBO formulation we just introduced can be solved in its current form by quantum annealers, but if we want to use a logic-based quantum computer, the problem needs to be reformulated somewhat. We can use the QAOA algorithm, which includes two important sets of actions to perform on your quantum computer.

One step (called a *mixing Hamiltonian*) randomly changes your routes a little (like taking a different turn at an intersection), and the other step (called a *problem Hamiltonian*) checks if the new route is shorter. You then repeat these two actions again and again. This sequence is called an *iteration*. With each iteration, you improve your guess for the shortest route. After a number of iterations, you measure the quantum state.

The translation from QUBO to QAOA is performed automatically by using the Strangeworks SDK QAOA extension.

Thus, when defining a QUBO problem to be inputted into the QAOA service, the QUBO problem is directly minimized, and the value of the cost function is displayed.

Asking Where Do We Go Now?

As Guns N' Roses sings in *Sweet Child 'O Mine*, "Where do we go? Where do we go now?" you've just run two quantum algorithms, but where do you go from here?

Well, you're already set up with an IDE, an SDK, and a quantum ecosystem with dozens of other applications for you to experiment with. So you can keep going for a long time in the Strangeworks environment if you choose to.

A ton of other resources are available. Check out Chapter 16 for educational resources to continue your quantum journey.

» Searching at super speed

» Finding new pharmaceuticals

Chapter 13

Quantum Computing Applications

The entire advantage of quantum computing is that it will execute certain specific computer algorithms much, much faster than the classical computers that people use today. We describe some of the key algorithms in the next chapter.

But even without diving into the details, we can describe the types of things that quantum computing will be very, very good at. And we can give a general idea as to which of these improvements might be available sooner rather than later.

So this chapter describes the application areas where quantum computing will make a difference. The next chapter describes several algorithms that will help drive quantum computing forward in the years ahead.

Thinking in Triplicate

There are three broad categories of quantum computing applications. It's useful to examine each task you're trying to accomplish from all three of these viewpoints. Applying quantum computing to real-world problems is a creative task, especially in these early days, and using multiple viewpoints can only be helpful.

Here are the three approaches:

- **Simulation:** In simulation, qubits — trapped bits of coherent matter, described in Chapters 10 and 11 — mimic other coherent matter, such as the individual atoms within a molecule that might become a medically useful drug. Simulation is arguably the most natural fit for quantum computing because quantum mechanics is what governs the laws of, well, nature.
- **Optimization:** A group of qubits can be used as a kind of computational furnace that can be guided into yielding a very good — but not necessarily perfect — solution to a problem. The result might be the right answer, or it may instead be something close to that. (A very good solution to a route-planning or investing problem might save, or make, you a lot of money, even if it isn't the best possible answer.)
- **Calculation:** This approach is, conceptually, the most like the classical computing problem-solving we're all used to. (The details, of course, are waaaaay different.) In calculation, qubits are combined into logic gates (see Chapter 10), making up a universal computer (see Chapter 2). When used as logic gates, qubits can solve any imaginable problem, and a quantum universal computer can solve some important problems far faster than today's computers — which also fit the "universal computer" description — but grind to a near-halt for some problems.

The original idea for quantum computers, from Paul Benioff and Richard Feynman, was for their use in simulating coherent matter so as to create new pharmaceuticals, materials, and so on (see Chapter 6).

Coherent matter is matter, usually at a very small scale — individual photons, electrons, and atoms, for instance — that is not interacting strongly with other matter and that is not being measured or observed, so that it obeys quantum mechanical principles, rather than classical mechanical principles. Think of a photon zipping out of the Solar System at the speed of light and you get the idea. Coherent matter can be in a state of superposition; can be entangled with other coherent matter; and can tunnel.

We can view the three categories of quantum computing applications as different types of maths problems (Cor blimey! They're onto the British usage!). Simulation requires solving differential equations; optimization requires combinatorial, well, optimization; and calculation requires solving complex problems in linear algebra and involves a lot of matrix math. Both the features used in machine learning and the operations against the Bloch sphere used for manipulating the qubits of gate-based quantum computers are stated as vectors, so the calculation approach is readily used for machine learning. (Although optimization can be used for machine learning as well.)

Algorithms can be grouped into these same three categories, which helps spotting areas where algorithms can be extended to accomplish additional goals. Importantly, the same quantum algorithm can underpin several different applications; for example, the algorithm that powers a financial portfolio optimization application might also underpin a separate application for route optimization.

Also, the categories of applications can overlap; for instance, if you use optimization to come up with better and better answers, you may at some point come up with the exact answer, just as if you used calculation. (For instance, using optimization to find the prime factors of a large prime number, just like Shor's algorithm, which belongs in the calculation category.) But the categories are useful for understanding the current state of quantum computing and anticipating what progress we might expect in the near future.

The potential for quantum computing is so exciting that we, your humble authors, and others involved in this field often say things like "quantum computing can crack RSA encryption." However, these statements are a kind of shorthand. They actually mean something like "RSA encryption is among the problems that we expect quantum computing to be able to solve." Unpacked, the statement really means something like the following: "Quantum computing demonstrates factorization capabilities and a rate of progress today that indicate it will be able to crack RSA encryption a decade or two in the future."

Cracking Cryptography with Quantum

Quantum cryptography is "the straw that stirs the drink" in quantum computing — a phrase first attributed to baseball great Reggie Jackson, who was working in an entirely different field (right field, to be precise). The current, fervent interest in quantum computing began in 1994 with the publication of Shor's algorithm (see Chapter 6), which is one of the few quantum algorithms that has been proven, at this early point, to have the potential for exponential speedup. However, Shor's algorithm will be able to do useful work only when it's run on quantum computers far more powerful than those available today.

Quantum computing has the potential to break the most common encryption methods used to secure digital communication today, such as RSA and ECC, which protect emails, bank information, the web, and more. These encryption methods rely on the difficulty of factoring large integers and the difficulty of computing discrete logarithms, respectively.

Quantum computers can perform these operations exponentially faster than classical computers, making them a threat to traditional encryption methods. Quantum algorithms have been proposed for key exchange, digital signatures, and encryption, which are the building blocks of secure communication.

When used for cybersecurity, quantum technology, like the Roman god Janus, has two faces:

- **Offensive (quantum computers):** Quantum computing can be used to crack many encryption schemes. Shor's algorithm will at some point be able to crack RSA encryption, the most common scheme used today. This scenario is terrifying to cybersecurity experts and, in particular, national security mavens, who started paying close attention to quantum computing in 1994 and have not stopped since.
- **Defensive (quantum cybersecurity):** Ideally, quantum cybersecurity will be used to encrypt data in ways that are far more powerful than today's encryption schemas and are able to withstand attacks from Shor's algorithm and other approaches. Two approaches being worked on now are called lattice-based cryptography and code-based cryptography. Essentially, those working on defensive technology are in a race to bring their solutions to market, and to foster widespread adoption, years before a quantum computer that can crack Shor's is developed. But information protected by today's standard encryption is likely to become accessible to quantum computing powered decryption a decade or two from now.

As we are finishing this book, Chinese researchers have demonstrated a network 1,000km (roughly 600 miles) long for unhackable quantum communications — the longest such network yet — using photons in fiber-optic cables. Because quantum particles, such as coherent photons, can't be copied or amplified, suppressing noise on the fiber-optic network used was crucial to this achievement.

For this experiment, the scientists used ultra-low-noise, superconducting, nanowire single-photon detectors to suppress system noise and used dual-band phase estimation. The study verified the feasibility of quantum key distribution at this distance, though even longer distances will be needed for deployment in production.

Quantum cybersecurity falls into two categories:

- **Hardware solutions:** Often referred to as quantum communication networks, which are essentially physical fiber-optic networks that entangled photons travel across. They are theoretically unhackable because observing the information in transit causes the information to collapse, preserving ultimate secrecy in transit. These networks, however, face current challenges in the distance that quantum information can travel, as well as the limitations of scaling due to the need for physical network infrastructure.

» **Software solutions:** Often referred to as post-quantum cryptography or quantum-resistant algorithms. These are algorithmic methods (think RSA), but the intent is to design them to be resistant to deciphering at the hands of a mature quantum computer. The National Institutes of Standards and Technology (NIST) in the US is working to certify a handful of these algorithms as secure against a mature quantum computer.

The reality is that most quantum cryptography (hardware) and post-quantum cryptography (software) solutions have a long way to go before they become viable commercial products, and it's likely that people will use these solutions with each other, depending on what needs to be secured.

One of the most promising algorithms in post-quantum cryptography is the area of lattice-based cryptography, which involves the use of mathematical structures called lattices to create cryptographic keys resistant to quantum attacks. Another algorithm gaining traction is code-based cryptography, which involves the use of error-correcting codes to create cryptographic keys. These codes are designed to be difficult to break, even by quantum computers.

So there is an arms race between offensive and defensive uses of quantum computing for cybersecurity. Unfortunately, even once quantum cybersecurity solutions are available, many experts believe adoption will take another ten years.

Let's say you've circulated information that you want to keep private for a long, long time. (The B-52 bomber, first flown in the 1950s, is expected to still be in service in the 2050s.) And you've protected key information about it with RSA.

All an adversary needs to do is get a copy of that RSA-protected information now, store it, and start working on quantum computing (or wait as others do so). At some point before 2050, they're likely to be able to crack that information. Worse, if the adversary develops that ability before you do, they will know your bomber secrets, and you won't know theirs.

This approach is called "harvest and decrypt," and it's the reason why developing quantum cybersecurity solutions is important to have *now*, as many years as possible before a quantum computer that can run Shor's algorithm is developed.

TECHNICAL STUFF

The day that a quantum computer can break RSA is referred to as *Q-Day*, and the capability of quantum computers to crack RSA and other encryption methods is called the *quantum apocalypse.* RSA comes in greater and lesser strengths, and the lesser strengths — which use much smaller prime numbers — will be vulnerable years before the greater strengths, so there won't be a single Q-Day; the quantum apocalypse will unfold over time. But these terms are useful for summing up the fear raised by offensive uses of quantum cybersecurity.

Fear of this capability has been, and continues to be, a big driver of activity in quantum computing, even if the general public doesn't know all the details.

Luckily, quantum computing doesn't spell only doom and gloom, and the promise of applications that will do things like cure cancer or reverse climate change are driving a lot of hope and balancing out the fearful narrative. These are the many uses of the technology that will make life better — in some cases, far better — than what people experience today. The rest of this chapter is devoted to these positive applications.

WARNING

Even positive applications of quantum computing can be used in neutral, negative, or criminal ways. For instance, the same kinds of drug development breakthroughs that might cure diseases could also be used to create hallucinogens or poisons. If you're in a position to guide your business or other organization into this future, consider the full range of possibilities in each area of technology.

QKD FOR THE WIN

An early and fundamental application of quantum cryptography is the use of quantum key distribution (QKD) as a means of establishing a completely secure communication channel between two parties. QKD operates by leveraging the unique properties of quantum mechanics to generate a shared secret key that is known by only the communicating parties.

This key is produced through the use of photons, which are transmitted between the parties. In the event that an unauthorized third party attempts to intercept the photons, the act of measuring them alters their state, thereby alerting the communicating parties to the presence of an eavesdropper. This enables the parties to take the appropriate measures to ensure the security of their communication.

QKD is being used today, though it's a very small part of the market. This initial use of QKD is likely to be only the first step in the growth of quantum communications as a real business.

Finding Waldo in a Sea of Striped Hats

Those of us who grew up on the *Where's Waldo* books (called *Where's Wally* in the UK, and by other names elsewhere) — or read them to kids of our own — will remember the complex crowd scenes that filled the books, with Waldo and his striped hat hidden in the crowd. The *Where's Waldo* books make people use their own image search and recognition abilities in a fun way.

Search algorithms have been an important area of research in computer science for decades. Real-world examples of the use of quantum algorithms for search include optimization problems in internet search, finance, logistics, and transportation. For example, the use of quantum algorithms for portfolio optimization will help financial analysts find the optimal investment strategy for a given portfolio in a fraction of the time required by classical algorithms. (Using quantum algorithms to optimize your portfolio works especially well if you have a quantum computer and the other investors don't.)

With the exponential growth of data, several algorithmic challenges need to be addressed. One of the biggest challenges is finding an optimal solution in a reasonable amount of time, which is where quantum algorithms come into play. One of the earliest, best-known, and most promising quantum algorithms is Grover's algorithm (see Chapter 14), used for searching an unsorted database and for a wide range of other purposes as well.

Another example of the use of a quantum algorithm for search is the quantum walk algorithm. This algorithm can search a graph or a network for a specific node or set of nodes. The algorithm uses the properties of quantum mechanics to traverse the graph or network in superposition, which allows it to search the entire graph or network in a fraction of the time required by classical algorithms.

QUANTUM-SAFE COMMUNICATIONS?

Researchers believe that quantum cryptography will be able to produce algorithms that are safe from decipherment by future quantum computing algorithms. However, researchers were also, for a long time, able to prove theoretically that hummingbird flight is impossible. But the little buggers continued to dart around, making a mockery of the relevant calculations. (It turns out that you have to include the Brownian motion of air molecules to understand hummingbird flight; you're welcome.) So all of this uncertainty leaves the practice of cryptography, quantum and non-quantum, in an unsettled state at present.

Grabbing That Cash

Before Waldo/Wally/et al, some of us grew up on the progressive rock of the 1970s and beyond. One song that still gets played today is the hit "Money" by Pink Floyd, which one of the authors — not the one you might think — is particularly attached to. (And our editor too!)

Quantum computing is starting to make waves in the financial industry, with many companies turning to this new technology in an effort to improve their operations and gain a competitive edge. Today, quantum algorithms and applications are being explored by a variety of financial companies for uses including portfolio optimization, risk management, and fraud detection.

Goldman Sachs, a leading investment bank, and several other banks are working to develop quantum algorithms for portfolio optimization; "the vampire squid," as Goldman Sachs is sometimes called, has shown promising results in improving investment returns. By utilizing the processing power of quantum computing, this portfolio optimization effectively analyzes vast amounts of data and identifies investment opportunities that traditional algorithms might overlook, leading to more informed investment decisions.

Other financial institutions are also exploring the potential of quantum algorithms in finance. JPMorgan Chase, for example, has been researching applications in portfolio optimization and options pricing. IBM has been partnering with several financial institutions to develop applications that can be utilized for risk management and fraud detection. (And this is just the work we know about!) The algorithms that power these applications are designed to take advantage of the unique properties of quantum computing, such as the capability to perform multiple calculations simultaneously, to provide more accurate and efficient solutions.

With the capability to simultaneously perform multiple calculations, quantum algorithms can help financial institutions make more informed decisions while minimizing risk and maximizing returns.

Looking to the future, researchers are exploring a range of new quantum algorithms that could change the way the financial industry does things even further. One area of focus is quantum machine learning, which uses quantum computing to train machine learning models more quickly and accurately than traditional methods. And quantum cryptography, mentioned earlier in this chapter, is of course of vital importance to the financial services industry.

Quantum algorithms have the potential to make the financial services industry more efficient, profitable, and secure. (And yes, we're sure it can be used for ethically dubious or illicit purposes as well.) While there is still much research to be done, it is highly likely that quantum computing will play an increasingly important role in finance in the years to come.

Insuring That Quantum Makes Its Mark

(Pun alert: *Ensuring* is the correct word for this purpose, but we deliberately used *Insuring* in the header as a pun.)

One area where quantum algorithms may be particularly useful in the insurance industry is in risk analysis. Insurance companies use risk analysis to determine the likelihood of a particular event occurring and the potential costs associated with that event. Quantum algorithms could greatly enhance this process by allowing for more complex calculations to be performed in a shorter amount of time. This, in turn, would allow insurance companies to better assess risk and set more accurate premiums.

Another area where quantum algorithms could be beneficial in the insurance industry is in fraud detection. Fraudulent claims cost insurance companies billions of dollars each year. Detecting and preventing fraud is a top priority for many insurers. Quantum algorithms could help insurers more effectively identify fraudulent claims by analyzing large amounts of data and detecting patterns that might be difficult to spot using traditional methods.

One such algorithm that is gaining momentum is Grover's algorithm (see Chapter 14), which can search through vast amounts of data in a fraction of the time that it would take traditional algorithms to do so. As the importance of data analysis continues to grow in the insurance industry, quantum algorithms are poised to play a significant role in streamlining operations and enhancing overall performance.

Several companies are already exploring the potential of quantum algorithms in insurance. For example, Allianz has partnered with IBM to develop a quantum computing platform that could be used for risk analysis and other applications. Similarly, Swiss Re has been researching the potential of quantum computing for several years and has already developed a prototype algorithm for calculating the value of financial derivatives.

In addition to these companies, many other insurance providers are likely to explore the potential of quantum algorithms in the years to come. For insurers that can successfully harness the power of quantum algorithms, the benefits could be significant, including improved risk assessment, more effective fraud detection, and greater efficiency in a variety of other areas.

Making the World Go Round with Logistics

Logistics is, in some ways, an odd industry. The term refers to the acquisition, movement, and storage of physical resources. Almost every business has some logistics aspect to it, but that can be as minor as bringing in office supplies for an accounting firm, or as resource-intensive as the need to move tons of steel by rail for use in manufacturing. So the logistics industry includes some things that almost every company does, as well as companies that specialize in logistics as their main business.

The logistics industry is constantly seeking ways to optimize its supply chain processes, and one of the latest innovations that has emerged is the use of quantum algorithms. Given the intricacies involved in supply chain optimization, quantum algorithms have the potential to be highly effective in this domain. They can facilitate the analysis of large data sets, optimize shipping routes, reduce transportation costs, and increase overall operational efficiency.

It's easy to assume that processing large data sets with quantum computing will require the use of quantum RAM or quantum hard disks. (Classical RAM and hard disks do take advantage of quantum effects but do not depend on them in the same fundamental way as a quantum computing version would.) But qubits are the only component we need to get through much larger data sets. In both classical and quantum computing, the data moves from RAM to the CPU or qubit for processing and is then returned to RAM. What's different with quantum computing is that, for the right operations, that processing step can be orders of magnitude faster than on a classical computer. And that means you can crunch far more data in a given period of time than on a classical computer, even without quantum RAM or quantum hard disks to help out. And yes, we are trying to imagine what those will be like when they do arrive, right along with all of you.

Figure 13-1 shows the optimization steps performed by quantum-inspired computing to help optimize logistics for Fujitsu and Toyota in Japan. Quantum computing excels in areas where combinations of options pile on top of each other, creating very large, multidimensional matrices of possibilities.

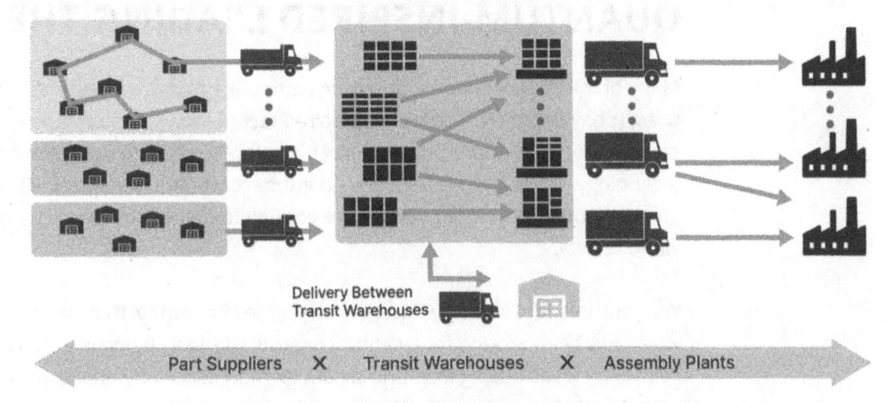

FIGURE 13-1: Quantum-inspired computing is being used by Fujitsu and Toyota to optimize logistics.

One easy-to-understand example of the power of logistics is the daily route planning used by delivery company UPS. They rather famously train their drivers, and design their routes, to almost always avoid turning left. This is not some kind of political statement, but rather the result of the long waits that drivers of all vehicles sometimes suffer in getting the opportunity to safely make a left turn. By avoiding them, UPS drivers save time and money. (And might even avoid a few bent fenders along the way.)

This "hack" makes the need for a powerful technology such as quantum computing clear. Every day, UPS and other delivery — and, yes, logistics — companies face the "traveling salesperson" problem of how to get the right packages, in the right trucks, with the right, right-turning (ha ha) drivers to save time and money in package delivery.

Fully optimizing this problem for a single day's deliveries in just one city might take all the computers on Earth many years to solve. Since no one has time for that, routes are far from optimal, and to help, one-size-fits-all approaches such as "don't turn left" are used. As quantum computers get more powerful, using them to route planes, trains, and automobiles will save ever-increasing amounts of time, energy, and money.

While there is still much research to be done in this field, it is clear that quantum computing and quantum algorithms will play an increasingly important role in logistics in the years to come. As such, companies such as DHL, FedEx, UPS, and even the USPS that look to remain competitive and stay ahead of the curve are likely to explore the potential benefits of these technologies and determine how best to incorporate them into their operations.

QUANTUM-INSPIRED LEADING THE WAY

Many of the applications for quantum computing that are being investigated, or even used, today depend on quantum-inspired technology (Chapter 8), which runs on classical computers, or quantum annealing (Chapter 9), a less complex form of quantum technology. While logic-gate-based quantum computing (Chapters 10 and 11) has the greatest potential, it is not yet mature enough to demonstrate much in the way of real-world results.

Your authors find it intriguing, and even somewhat funny, that classical computing, which has been around for a century or so, is suddenly being inspired to reach new heights by ideas arising from the challenge represented by quantum computing, an entirely new technology. It was okay to do things inefficiently for a century or so, and now it's all gonna get fixed so quantum computers don't get bragging rights?

At this writing, one of the hottest stocks with a quantum computing story to tell is Nvidia (see Chapter 15), which makes classical CPUs and GPUs (graphics processing units, which are very fast at certain kinds of math, including math used in AI). Nvidia, at present, has zero involvement in qubits or quantum annealing, but their chips are very useful for running quantum simulators and for use in quantum-inspired computing.

Dreaming of Machines Learning

The use of quantum algorithms for machine learning and artificial intelligence may be the most exciting prospect of all. Quantum computers offer the potential to solve problems related to the use of machine learning for optimization, simulation, pattern recognition, and more.

Several companies and research institutions are exploring the use of quantum algorithms in machine learning and artificial intelligence. Examples include the following:

- IBM and its IBM Q quantum computing platform, which offers access to quantum computing resources to researchers and developers. IBM has been working on developing quantum algorithms for machine learning, including the development of a quantum version of the well-known machine learning algorithm called support vector machines (SVM).
- Google has developed a quantum version of the neural network algorithm called the quantum neural network.

» Microsoft has used its Azure Quantum platform to develop a quantum version of the decision tree algorithm.

» Also involved are D-Wave Systems, Rigetti Computing, and Xanadu. For instance, Xanadu has been working on the development of a quantum version of a clustering algorithm.

The quantum support vector machine (QSVM) algorithm is designed to classify data by finding the optimal hyperplane that separates different classes of data. QSVM is faster than classical support vector machines, making it ideal for use in large data sets. Another example of a quantum algorithm for use in this space (no pun intended) is the quantum k-nearest neighbor (QKNN) algorithm, which can be used to classify data by finding the nearest neighbors to a given point in a high-dimensional space.

Searching for the New Oil in Quantum

It's long been said that "data is the new oil." Quantum computing can help us find better answers when data threatens to overwhelm us.

One important area where quantum algorithms show promise is in the field of optimization. Optimization, as we have mentioned, is the process of finding the best solution to a problem given a set of constraints. Many real-world problems, such as scheduling and logistics, can be modeled as optimization problems. Classical computers struggle to solve these problems efficiently, but quantum algorithms offer the potential for exponential speedup.

One example of a quantum optimization algorithm is quantum annealing, which is used to find the global minimum of a function by slowly cooling — reducing the energy available to — a system of qubits. This algorithm has been used to solve optimization problems in finance, logistics, and drug discovery.

Another example is the quantum approximate optimization algorithm (QAOA), designed to find approximate solutions to optimization problems. QAOA has been used to optimize the placement of cellphone towers in a network, as well as to optimize the routing of delivery vehicles for a logistics company. (Just to be clear, as of today, the vast majority of delivery vehicle routes and cellphone tower placements are calculated on classical computers.)

Real-world examples of companies using quantum algorithms in their data science efforts are still relatively few and far between, but there are some notable examples. D-Wave Systems (see Chapter 9) has developed a quantum annealing machine used by several large companies, including Volkswagen, Lockheed Martin, and Google. Volkswagen is using the D-Wave machine to optimize traffic flow in Beijing, while Lockheed Martin is using their access to optimize the design of aircraft components. Google is using the D-Wave machine to explore machine learning applications of quantum computing.

Making Materialism Matter

Quantum algorithms are emerging as a powerful tool in material science. One of the most promising quantum algorithms for material science is the quantum chemistry algorithm. This algorithm is designed to simulate the behavior of molecules and materials at the quantum level, enabling researchers to predict their properties and behavior with high accuracy. Using quantum chemistry algorithms, researchers can simulate chemical reactions, understand the behavior of materials under extreme conditions, and design new materials with specific properties.

Quantum annealing is also useful in materials science because the algorithms can help researchers find new materials with specific properties, such as superconducting capability or high strength-to-weight ratios. D-Wave has developed a quantum annealing algorithm designed for materials science research and has partnered with several universities and research institutions to explore the potential of this technology.

IBM has developed a quantum chemistry algorithm designed to simulate the behavior of molecules and materials at the quantum level. This algorithm has already been used to predict the behavior of several important molecules, including caffeine and carbon dioxide, and is expected to have a major effect on materials science research in the future.

Quantum computing and machine learning have been used at the University of Notre Dame to create glass that serves as a kind of shade (see Figure 13-2), deflecting heat and thereby lowering cooling costs.

In addition to these companies, several research groups are exploring the potential of quantum algorithms for material science. For example, researchers at the University of California, Berkeley, use quantum annealing algorithms to design new materials with specific properties, such as materials that can store energy more efficiently. Similarly, researchers at the University of Oxford are using quantum chemistry algorithms to study the behavior of catalysts, which are critical for many industrial processes.

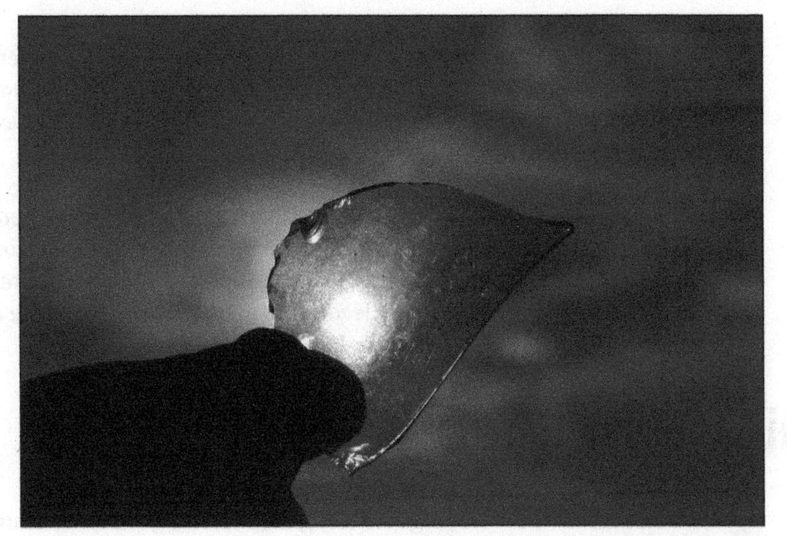

FIGURE 13-2: Quantum computing and machine learning are being used to create glass that deflects heat and lowers air-conditioning costs.

Denis Lytiagin / Adobe Stock

Simulating Our Way to Better Health

One of the most promising applications of quantum algorithms in medical science is in modeling the workings of the human body at the molecular level. Quantum computers can succeed here where classical computers fall short.

One real-world example of the use of quantum algorithms is the work being done by researchers at the University of Toronto. They have used quantum algorithms to simulate the behavior of a protein involved in the development of cancer. By doing so, they were able to identify a potential drug candidate that could inhibit the protein's activity, potentially leading to new cancer treatments.

Another area where quantum algorithms are showing promise is in medical imaging. MRI scans, for example, produce vast amounts of data that must be processed and analyzed to produce images of the body. Classical computers can struggle with this task, but quantum algorithms can handle it much more efficiently, which could lead to faster and more accurate diagnoses, as well as more effective treatments.

One notable example is the work being done by researchers at IBM. They have developed a quantum algorithm that can analyze the complex data produced by MRI scans, allowing doctors to identify tumors and other abnormalities more accurately. This research could lead to earlier detection and treatment of diseases, saving lives.

Finally, quantum algorithms are used also to improve our understanding of biological systems. By simulating the behavior of complex biological systems, researchers can gain new insights into how they work and develop new treatments for diseases.

Researchers at the University of Southern California have used quantum algorithms to simulate the behavior of a protein involved in Parkinson's disease. They were able to identify a potential drug candidate that could prevent the protein from aggregating, potentially leading to new treatments for the disease.

Finding New Pharmaceuticals

The process of developing new drugs is incredibly time-consuming and expensive, with many potential candidates failing in clinical trials. However, quantum algorithms can simulate the behavior of molecules at a level of detail that's impossible for classical computers. The effectiveness of quantum computers for this purpose means that researchers will be able to more accurately predict the effectiveness of different compounds, potentially leading to faster and more successful drug development.

One of the quantum algorithms being tried for drug discovery is the variational quantum eigensolver (VQE). This algorithm is used to determine the ground state energy of molecules, which is a critical factor in drug design.

The VQE algorithm uses a hybrid approach that combines classical and quantum computing to solve complex problems. It's particularly useful in drug discovery because it can accurately predict the molecular structure of compounds and their interactions with target proteins.

Another quantum algorithm that has gained traction in drug discovery is the QAOA algorithm we mentioned previously. It solves optimization problems, which are common in drug discovery. The QAOA algorithm uses a series of quantum gates to optimize the energy landscape of molecules, which helps researchers identify the most promising drug candidates.

One real-world example of the successful use of quantum algorithms in drug discovery is the work of Zapata Computing. This company has developed a quantum computing platform that uses the VQE algorithm to accelerate the drug discovery process. Their platform has been used as part of the effort to design new drugs for cancer and other diseases, and they have reported significant improvements in accuracy and efficiency.

Another company that is using quantum algorithms for drug discovery is Cambridge Quantum Computing. They have developed a platform that uses the QAOA algorithm to optimize the molecular structure of compounds. Their platform has been used to design new drugs for Alzheimer's disease and other neurological disorders.

Forecasting Future Fog

As we continue to learn more about the world we live in, one area of study that has become increasingly important is the field of weather forecasting and climate change studies. With the help of quantum algorithms, scientists and researchers can gather more accurate data and make more informed predictions about the weather and its effect on our planet.

One of the most well-known quantum algorithms used in weather forecasting is the quantum Monte Carlo algorithm. (It was not originally a quantum computing algorithm, but researchers are now trying a version on quantum computers.) This algorithm allows scientists to simulate the behavior of particles in a system, which can be useful for predicting weather patterns. It has been used by companies such as IBM and Google to improve the accuracy of weather-forecasting models.

THE CLIMATE IT IS A-CHANGING

Climate change is a looming crisis that requires innovative solutions. The use of quantum computing and quantum algorithms could be one such solution. These technologies can help us better understand climate patterns and predict future climate changes with greater accuracy. By simulating complex systems and performing calculations at a much faster rate, quantum algorithms could help us identify ways to reduce carbon emissions, trap carbon from manufacturing processes or in ambient air, and develop more efficient renewable energy sources.

With the potential to revolutionize our approach to addressing climate change, quantum computing and quantum algorithms offer hope for a more sustainable future. One of the authors (whurley) received an Eisenhower Fellowship for his climate-related project, Qlimate. . . yes, with a Q. (Before he was president of the United States, Dwight D. Eisenhower led the 1943 D-Day invasion of Europe in World War II. The Eisenhower Fellowship supports youthful leaders in "invading" one another's countries, but "armed" only with new ideas, PowerPoint presentations, thesis proposals, and other tools of peaceful and productive discussion.)

A commonly used quantum algorithm in weather forecasting is Grover's algorithm, the sorting algorithm mentioned earlier in this chapter. The capability of Grover's algorithm to search through large amounts of data quickly and efficiently makes it useful for analyzing weather data and identifying trends. It has been used by D-Wave Systems to improve the accuracy of climate change models.

The quantum annealing algorithm's capability to optimize complex problems can be useful for predicting weather patterns. Lockheed Martin has used it to help improve the accuracy of weather-forecasting models.

One real-world use case of quantum algorithms in weather forecasting is in the field of hurricane prediction. Researchers at the University of Maryland have used quantum annealing to predict the paths of hurricanes, which can help emergency responders prepare for potential disasters. Another real-world use case is in the field of aviation, where quantum algorithms are used to improve the accuracy of weather forecasting models, which can help airlines avoid dangerous weather conditions.

Overall, the use of quantum algorithms in weather forecasting and climate change studies is proving to be a valuable tool for scientists and researchers. With continued advancements in quantum computing technology, we can expect to see even more progress in this field in the years to come.

» Taking a trip to the quantum zoo

» Making Shor we can crack all the codes

» Seeing what Simon says about periodicity

Chapter 14

Quantum Computing Algorithms

Quantum computing algorithms have been a topic of intense research in recent years due to their crucial role in enabling quantum computers to solve problems that classical computers are unable to complete — or at least, not in a reasonable amount of time.

The word *algorithm* comes from the name of a ninth-century Persian mathematician, Muhammad ibn Mūsā al'Khwārizm. The Latinized version of his name, Algoritmi, was used for centuries to refer to "the decimal number system" — but with quantum algorithms, we go well beyond decimal numbers.

Although the development of quantum algorithms is still in its early stages, notable progress has been made, and several quantum algorithms have already been proposed. These algorithms have the potential to revolutionize several fields, including cryptography, optimization, and pharmaceuticals, as described in Chapter 13.

TIP

The contents of this chapter will be too advanced for many readers to follow closely, and only those who are already spending time with quantum computer algorithms will be fully conversant with it. But we recommend that every reader at least give this chapter a look; you'll begin to pick up some of the vocabulary that will help you follow future advances in the field.

Reviewing the algorithms here will give you a sense of what quantum computing will be able to accomplish as quantum computers become more powerful in the years ahead. It can also help you tackle new problems. Both existing and new problems may require a mix of one or more existing algorithms plus new work to tackle problems that could not have been solved, or that could only have been solved in significantly more time and at greater cost, on classical computers.

REMEMBER

An algorithm is a series of steps that are followed to complete a calculation or other problem-solving operation. Quantum computer algorithms have one or more steps that benefit from being executed on a quantum computer or quantum-inspired hardware.

Mapping Quantum Computing Algorithms to Applications

Table 14-1 shows how quantum computing algorithms (described in this chapter) map to application areas (discussed in Chapter 13). Expect the situation to get more complicated as new algorithms are created and additional uses are found for existing algorithms.

TABLE 14-1: Some Quantum Computing Algorithms Are Useful across a Range of Application Areas

	Cryptography	Search and Optimization	Machine Learning
Shor's algorithm	x		
Grover's algorithm		x	
Quantum phase estimation (QPE) algorithm		x	
Simon's algorithm	x		
Quantum Fourier transform (QFT) algorithm		x	

	Cryptography	Search and Optimization	Machine Learning
Harro-Hassidim-Lloyd (HHL) algorithm		x	
Quantum approximate optimization algorithm (QAOA)		x	
Variational quantum eigensolver (VQE) algorithm		x	
Quantum counting algorithm		x	
Quantum Fourier sampling algorithm		x	
Quantum phase estimation with a quantum neural network (QPE-QNN) algorithm			x
Amplitude estimation algorithm		x	
Quantum principal component analysis (PCA) algorithm			x
Quantum support vector machine (QSVM) algorithm			x
Quantum walks algorithm		x	
Hidden subgroup problem (HSP) algorithms		x	
Quantum matrix inversion algorithm		x	
Quantum k-means algorithm			x
Quantum approximation of nonlinear functions (QANF) algorithm		x	

Quantum algorithms are interrelated in interesting ways. Figure 14-1 shows a kind of family tree of quantum computing algorithms. As you can see from the figure, if you learn just a few core quantum algorithms shown in the figure — quantum Fourier transform (QFT) on the left side of the figure and quantum walks and graph algorithms on the right side — you unlock the core of most quantum computing algorithms, and you can go apply your newfound understanding to similar problems in your own work.

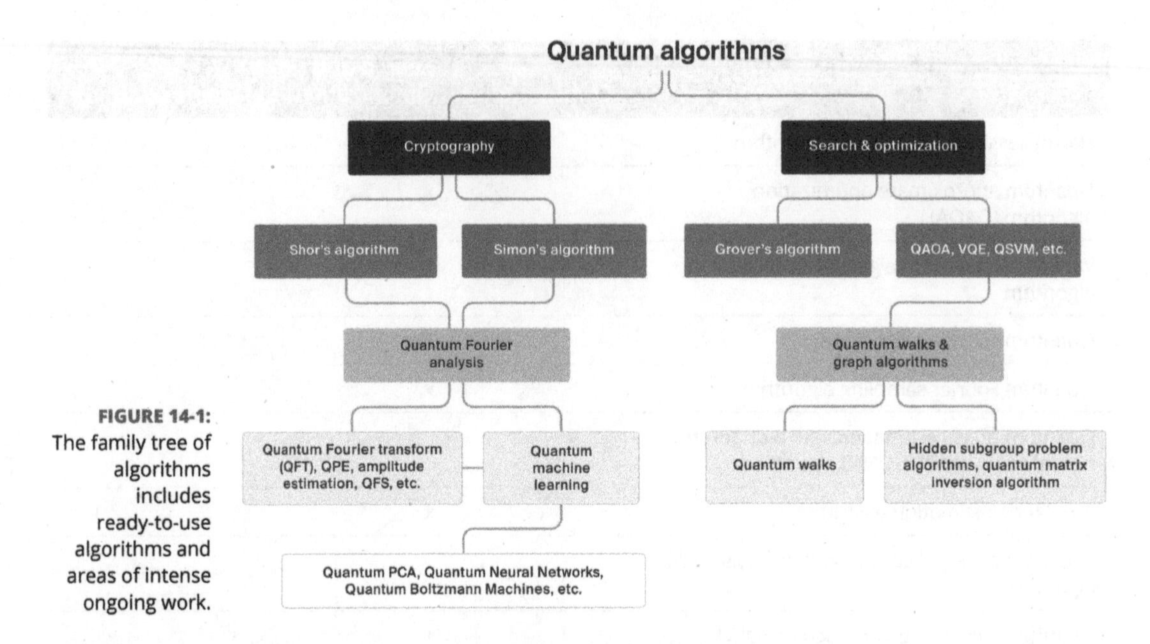

FIGURE 14-1: The family tree of algorithms includes ready-to-use algorithms and areas of intense ongoing work.

Understanding the Basics of Quantum Algorithms

Quantum algorithms are designed to harness the unique properties of quantum mechanics to solve complex problems more efficiently than classical algorithms. Quantum algorithms manipulate quantum bits, or qubits, which can exist in multiple states simultaneously. (See Chapters 10 and 11 for specifics.) Qubits can also be entangled with each other, with the two capabilities together allowing for parallel computation.

Quantum algorithms include specific steps that can be executed usefully on a quantum computer, with its quantum-mechanics-enabled capability to bring back the right answer from among an immense number of possibilities. Other steps in the same algorithm might execute just as well, or better, on a classical computer. Quantum computing will include a lot of classical computing as part of the deal.

Mapping algorithms to use cases

The exploration of quantum algorithms is steadily moving out of academic labs and into the innovation centers of Global 2000 companies. The real-world research focus that business interests can bring is critical for advancing quantum algorithms to practical usage, since their work often leads to improvements in the algorithms. This work also delivers a deeper understanding of the quantum computing hardware improvements needed to make the quantum algorithm ready for commercial use.

The same quantum algorithm may be applied in many different areas. These and other quantum algorithms are being explored and implemented in various fields, from finance to medicine, and are expected to revolutionize the way we approach complex problems.

Let's explore quantum algorithms in more depth, taking a closer look at some specific quantum algorithms that have sparked interest and excitement in the field of quantum computing. We hope this will provide you with a high-level understanding of these algorithms and their potential applications. Whether you are a beginner or an experienced quantum computing enthusiast, we believe this to be one of the most valuable actions in this book. So, let's dive in and explore the fascinating world of quantum algorithms together!

Investing in research

Governments around the world have invested significant billions of dollars into quantum research in recent years, with large national efforts in the US, China, Canada, the UK, Australia, Germany, and more. In many cases, government investment has been accompanied by academic focus and significant business investments. These investments have led to significant advances in the field, with many researchers now working on developing more powerful and reliable quantum hardware and software.

However, as hardware capabilities improve, research on quantum algorithms is expected to increase. While hardware development will still be vitally important, many researchers believe that the real breakthroughs in quantum technology will come from developing new algorithms that can take advantage of the unique capabilities of quantum computers.

QUANTUM COMPUTING IS NOT JUST ON QUBITS

Unlocking the power of qubits is the moon shot of quantum computing. However, quantum simulators and quantum-inspired computing (see Chapter 8), all of which run on classical computing hardware, are also areas of active investigation and work. In addition, efficient burden sharing between classical computers and quantum computers is needed to solve most interesting problems, and optimizing this sharing continues to drive work in hardware and software, including in algorithms.

This shift in focus is expected to be driven by a number of factors, including the increasing complexity of quantum hardware and the growing need for new and more powerful algorithms to solve increasingly complex problems. As a result, governments are likely to invest more money into quantum algorithm development in the coming years, with a particular focus on applications in areas such as finance, logistics, and healthcare. Ultimately, the hope is that all of these investments will lead to a new era of computing, where quantum technology is used to solve some of the world's most pressing problems.

HOW MUCH HAVE GOVERNMENTS INVESTED?

Governments around the world are investing heavily in quantum computing research and development. According to a report by the World Economic Forum, government expenditures in one recent year were estimated at roughly $2B.

In the United States, the federal government has allocated $1.2 billion over the next five years for quantum research through the National Quantum Initiative Act. China has dedicated considerable resources and attention to quantum technologies, having announced the intent to invest $10 billion in a national quantum research center. The European Union has committed €1 billion to support quantum research and development initiatives, and the UK has committed £2.5 billion. Private industry is spending additional billions.

These are just the public announcements. The comedian Steve Martin once complained about the award-winning martial arts movie, *Crouching Tiger, Hidden Dragon* (2000), because it didn't show any tigers or dragons. It was explained to him that, as the title describes, the tigers were crouching and the dragons were hidden. We don't *think* he was talking about quantum computing when he made this joke.

In a similar vein, the funding announcements we describe here are probably like icebergs — what you see (the part above the waterline) is not what you get (the total mass of the iceberg). And remember that the *Titanic* was not sunk by the visible part of the iceberg.

However, even the sums announced publicly to date demonstrate the global recognition of the potential effect of quantum computing on various industries and the need to stay competitive in this emerging field.

Visiting the Quantum Zoo

Stephen Jordan from Microsoft Quantum created and maintains Quantum Algorithm Zoo, a comprehensive catalog of quantum algorithms. The Quantum Algorithm Zoo, shown in Figure 14-2, breaks algorithms into categories including Algebraic & Number Theoretic, Oracular, Approximation and Simulation, Optimization, Numerics, and Machine Learning. While this would be considered advanced material for most people, it's fun and educational to browse through the algorithms and their descriptions. The Zoo is a useful resource as your understanding of quantum computing algorithms grows.

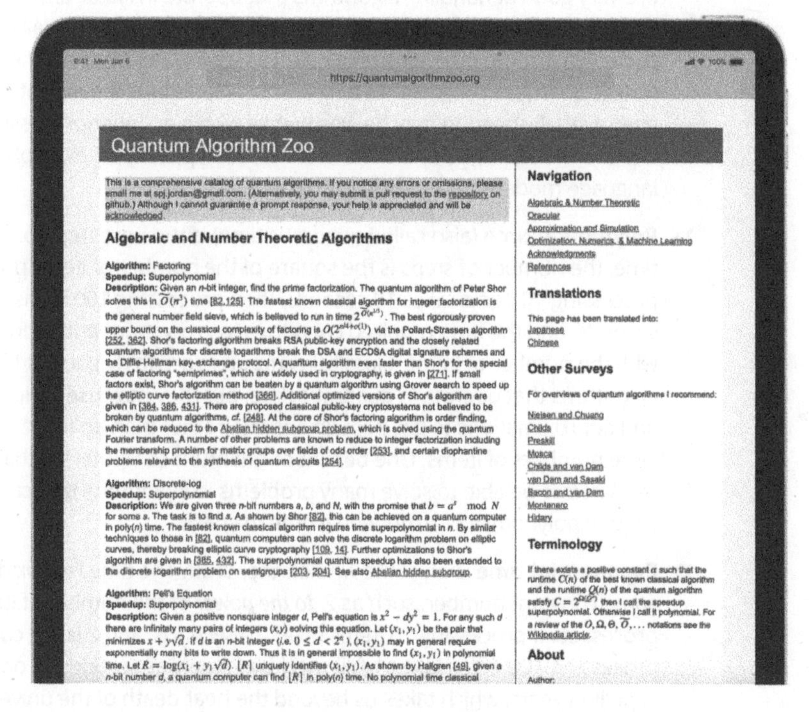

FIGURE 14-2: The Quantum Algorithm Zoo is a rich resource for work in quantum computing.

Finding New Kinds of Time

To understand how quantum computing algorithms make computing faster, you need a different way of looking at time. Yes, you, too, can be a Time Lord — a member of the ancient, yet modern extraterrestrial race from *Doctor Who*. The Time Lords have mastered time travel and view time in a nonlinear way.

The *time warp* is entirely different; that's the theme song from *The Rocky Horror Picture Show* (1975). Let's do the time warp again — and again and again, if you're a Time Lord as well.

In discussing computer algorithms, we divide them into categories based on how the execution time of the algorithm scales with the number of items to be processed:

- **Linear time** is a simple multiple of the number of items processed. If an algorithm operates in linear time, if processing 10 items takes 10 seconds, then processing 100 items takes 100 seconds, and so on. Classical computers are very good at handling algorithms that operate in linear time, because they — wait for it — operate in linear time themselves. It's usually not worth it to put linear-time problems on a quantum computer because there's no way for the quantum computer to deliver much speedup. The exception is that even a small speedup may be worthwhile within an inner processing loop that is performed millions or billions of times, as happens, for example, with large language models (LLMs) such as ChatGPT.
- **Polynomial time** (also called quadratic time) is the next step up. In quadratic time, the number of steps is the square of the number of items processed. If processing 10 items takes 10 seconds, then processing 100 items takes 1,000 seconds, and the time required continues to grow more and more quickly with the number of items. Classical computers are fairly good at handling algorithms that operate in quadratic time, but they may use various shortcuts and approximations to get a rough answer, rather than an exact one, for truly large numbers of items. One benefit of quantum computers is that they're expected to be able to solve many problems in linear time that currently require polynomial or exponential time to solve.
- **Exponential time** is on some whole other stuff. The time required here is some constant number, such as 2, *to the power of* the number of items processed. If processing 10 items takes 10 seconds, and 2 is the constant, processing 100 items takes 1-followed-by-29-zeroes seconds — or about a septillion years, which takes us beyond the heat death of the universe. Classical computers roll over, stick up their legs, and die on exponential-time algorithms with large numbers of items to process. Powerful future quantum computers will solve exponential-time problems in quadratic or linear time, unlocking a speedup that will result in solutions to some of the most challenging computational problems that humanity faces.

You will also see references to logarithmic time. In *logarithmic time*, the first cut at a problem may take a while, but successive steps take progressively less time. (Please send us a Zeno's paradox joke to use here.) For our purposes, logarithmic time is similar to linear time.

To sum up, linear time algorithms execute in multiples of n, the number of item processed; polynomial time algorithms execute in n raised to the power of a constant exponent (that is, n^2); exponential time algorithms execute in a constant base raised to the power of n (that is, 2^n).

REMEMBER

The main benefit of quantum computing is speed. As quantum computers become more powerful in the years ahead, they will be able to execute, in linear time, specific algorithms that execute in polynomial time on classical computers. And they will be able to execute in either polynomial or linear time, depending on the algorithm, specific algorithms that execute in exponential time on classical computers. As quantum computer power continues to increase further, they will be able to deliver these speedups for greater and greater values of n (that is, they will be able to solve more and more complex problems).

Figure 14-3 shows a chart that conveys the difference between the different kinds of algorithmic time, when algorithms are processed by classical computers. Quantum computing converts the extreme curves to less extreme ones.

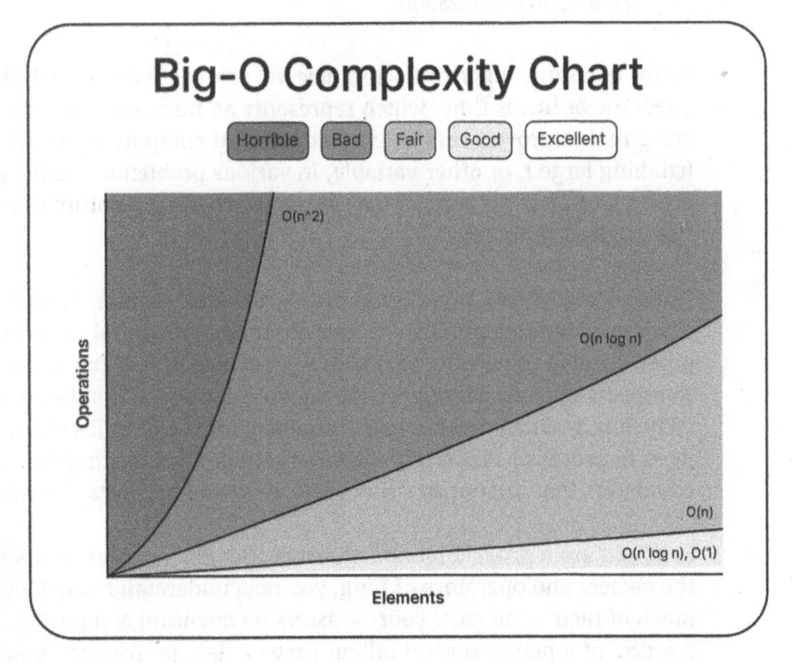

FIGURE 14-3: Time still has meaning in quantum computing — just a different meaning.

TECHNICAL STUFF

There's special notation for each kind of time. $O(n)$ denotes linear time; $O(n^2)$ denotes quadratic time; and $O(C^n)$ denotes exponential time. When you see the $O(C^n)$ notation, go looking for the latest and greatest quantum computer you can find, or we're all going to be here for a while.

These different kinds of time are used to describe the kind of speedup that quantum computing is expected to deliver when solving different kinds of problems:

- A *polynomial speedup* is when a classical computer requires an amount of time on the order of t^2 to solve a problem, but a quantum computer requires only on the order of time t. In this case, the time needed by the quantum computer is on the order of the square root of the time needed by a classical computer. Grover's search algorithm delivers a polynomial speedup; a search that requires roughly one million steps on a classical computer needs roughly one thousand steps on a quantum computer powerful enough to tackle this problem.
- An *exponential speedup* is when a classical computer takes on the order of time 2^t to solve the problem — which is a vastly larger amount of time than t^2, for large t — but a quantum computer needs only time t. For instance, if a classical computer needs 2^{100} seconds to solve a problem — the septillion years mentioned previously — a sufficiently powerful quantum computer needs on the order of 100 seconds. Shor's algorithm delivers an exponential speedup in factorization.

Many quantum algorithms promise to process exponential-time problems in quadratic or linear time, which represents an immense speedup. However, delivering real improvements over what classical computers can do today — that is, handling large t, or other variable, in various problems — will require very powerful quantum computers; it may be decades before quantum computers can tackle the toughest problems.

Grover's algorithm, mentioned in the previous section, "only" provides a quadratic speedup to a problem that would take exponential time on a classical computer: Instead of needing to process an average of half the items in a list, it needs to process only an average of the square root of the number of items in the list. (Which is a much, much smaller number, for the long lists that Google and Bing need to process.) Pleasingly, however, Grover's algorithm may run on quantum computers that are not so many years away as for Shor's algorithm.

So if you are a shareholder of Alphabet, Google's parent company, or Microsoft, the owners and operators of Bing, you now understand why they are investing so much of their — in part, your — assets on quantum computing. If you can save a fraction of a penny several billion times a day, pretty soon you're talking about real money, as well as an exponential (which is a pun, yes, but we mean it) advantage over the competition.

Starting with the Deutsch-Jozsa Algorithm

The *Deutsch-Jozsa algorithm* was invented by David Deutsch and Richard Jozsa in 1992. It's a problem that aims to distinguish between two types of functions. The first type of function is *constant*, meaning it has the same output for all inputs. The second type of function is *balanced*, meaning it has an equal number of 0s and 1s in its output. The problem is to determine which type of function a given function is.

The Deutsch-Jozsa algorithm can solve this problem in a single query to the function, whereas a classical algorithm requires at least two queries. This speedup is significant mostly because it's an early example showing that quantum computers can offer a significant advantage over classical computers for certain problems.

The Deutsch-Jozsa algorithm consists of four steps:

1. Prepare the input state, which is a superposition of all possible input values.
2. Apply a quantum gate (or series of gates) called the oracle to the input state. The oracle performs a transformation on the input state that depends on the function being evaluated.
3. Apply a second quantum gate to the input state. This gate, called the Hadamard gate, creates a new superposition of all possible output values.
4. Measure the output state and use the result to determine whether the function is constant or balanced.

The key to this algorithm is the use of the oracle. The *oracle* is a quantum gate that performs a transformation on the input state that depends on the function being evaluated. The transformation is such that it flips the phase of the output if the function is balanced but leaves the output unchanged if the function is constant. This transformation is what allows the algorithm to distinguish between the two types of functions.

The algorithm has several applications in function evaluation and decision problems. In function evaluation, the algorithm can be used to determine whether a function is constant or balanced. This capability can be useful in cryptography, where it's important to distinguish between secure and insecure encryption algorithms.

In decision problems, the algorithm can be used to make decisions based on the output of a function. For example, it can be used to determine whether a given set of data is malicious or benign.

Making Shor Quantum Computing Will Be Big

Shor's algorithm sounds innocent enough. It's a way to factor specific kinds of large numbers into their prime factors.

Start with two prime numbers — numbers that have no other factors except 1 and the number itself. 1, 3, 5, and 7 are all prime numbers; 9 is not, because it's the product of 3 and 3. No even number is a prime number, because is, by definition, divisible by 2. And as you count higher and higher, the number of prime numbers diminishes because there are so many smaller numbers to serve as potential factors.

As numbers get larger, it gets hard for computers to determine whether a given number is prime or not, and to determine its factors. Numbers don't have to be all that large before classical computers can't factor them (or determine that they are prime) in a reasonable amount of time — less than, say, several years.

As a result of this complexity, prime numbers are used in cryptography. A large number that's the product of two primes is used to encrypt information. If we give you one of the primes, called the secret key, it's easy for you to factor the number and access the information. But someone who has only the large number — the public key — can't easily break it down and access the information.

Unless, that is, they have two things: Shor's algorithm, which uses quantum computing to factor numbers much more quickly, and a quantum computer with many thousands of error-free qubits to power Shor's. Fortunately for the secret-keepers of the world — which is all of us, because our banking information is part of the secrets being kept — those quantum computers do not exist yet, and may still be 10 or 20 years away.

But just the threat that this combination may soon exist has riveted the attention of important players on quantum computing since Shor's algorithm was developed, as described in Chapter 6.

How does Shor's algorithm work? It's just a few steps — and only Step 4 requires quantum computing:

1. Choose a number that you want to factor. Let's say we want to factor the number 15.
2. Choose a random number between 1 and the number you're trying to factor. Let's say we choose the number 4.

3. Calculate the greatest common divisor (GCD) of your random number and the number you're trying to factor. The GCD is the biggest number that divides both of those numbers evenly. In our example, the GCD of 4 and 15 is 1. If the GCD is 1, move on to the next step. If it's not 1, you've already found one of the prime factors of your original number! You can stop here.
4. Here's where quantum computing comes in. In Shor's algorithm, you use it to figure out the period of a function that's applied to the target number. The *period* is like a repeating pattern that the function follows. This is the step where Shor's algorithm really shines because it can do this really quickly even for really big numbers.
5. Once we know the period of our function, more math tricks can be used (on a classical computer) to figure out the prime factors of our original number. In our example of 15, the prime factors are 3 and 5.

So that's Shor's Algorithm in a nutshell. It's being optimized even as we speak, in hopes that it can be used sooner rather than later — even though using it will cause a lot of trouble.

Searching with Grover

The *search problem* is a common problem in computer science. It involves finding a specific item in a large database, with the complexity of this problem increasing as the size of the database increases. Search techniques can also often be used for optimization problems.

Grover's algorithm is a quantum algorithm designed to find a specific item in an unsorted database. Grover's algorithm has a quadratic speedup over classical algorithms, making it an important tool for quantum computing. Classical algorithms that solve this problem have a complexity of $O(N)$, where N is the size of the database. However, Grover's algorithm has a complexity of O(sqrt(N)), which is a significant improvement in efficiency.

TECHNICAL STUFF

As sufficiently powerful quantum computers are developed, Grover's algorithm will deliver a speedup on the order of 10x for a list of 100 items; 1,000x for a list of a million items; and 1,000,000x for a list of a trillion items, which is not an incredibly long list to be searching these days. To wit, it will make previously unwieldy searches fast, and previously impractical searches fast enough to be practical.

Grover's algorithm works by using quantum parallelism to search for the needed item within the database. As with most quantum computing algorithms, the algorithm starts by initializing the quantum computer to a superposition of all possible states. The algorithm then applies a series of operations that increase the amplitude of the state that contains the desired item. This process is repeated for as many times as needed until the desired item is found.

The step-by-step breakdown of Grover's algorithm can be summarized as follows:

1. Initialize the quantum computer to a superposition of all possible states.
2. Apply the oracle function to mark the state that contains the desired item.
3. Apply the inversion about the mean operation to amplify the amplitude of the marked state.
4. Repeat Steps 2 and 3 a certain number of times until the desired item is found.

The oracle in a quantum computing algorithm is like a CASE statement in classical computing. It causes the algorithm to exhibit different behavior based on the value of a variable. The way in which it does this is specific to quantum computing and is, as the name implies, a bit magical — yet sensible, once you understand how the logic gates in a quantum computer work.

In optimization, Grover's algorithm can be used to find the minimum or maximum value of a function. This algorithm can optimize complex systems, such as financial portfolios or supply chain management.

Using the Quantum Phase Estimation Algorithm

It turns out that *phase estimation* is an essential problem in quantum computing because it is a crucial step in several quantum algorithms. A quantum state can be represented as a superposition of two or more states, where each state has a phase angle. The phase estimation problem is to compute the unknown phase angle of a given quantum state.

The phase estimation problem is challenging because quantum states are complex numbers and their phases are not directly observable. Therefore, quantum algorithms are required to estimate the phase angles. One of the most commonly used algorithms for phase estimation is the quantum phase estimation algorithm.

This algorithm uses the quantum Fourier transform to estimate the phase angle of a quantum state. The algorithm is composed of two parts: a preparatory part and an iterative part.

The preparatory part involves preparing the input state, which is a superposition of the state whose phase angle needs to be estimated and an ancillary qubit. The ancillary qubit is used to store the estimated phase angle. The input state is then transformed into a state that can be used for the iterative part of the algorithm.

The iterative part involves applying the quantum Fourier transform to the input state. The quantum Fourier transform is a quantum algorithm that maps a quantum state to its Fourier coefficients. The Fourier coefficients of the input state are then measured, and the estimated phase angle is obtained from the measurement results.

Figure 14-4 shows a Bloch sphere of the type used to picture the transformations that quantum computers apply to qubits as they calculate results.

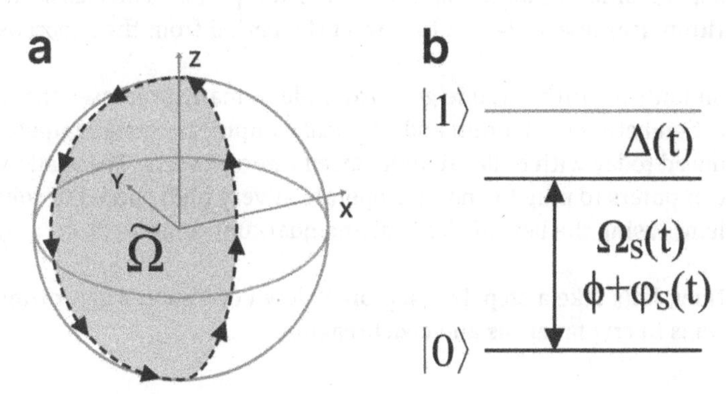

FIGURE 14-4: The quantum phase estimation algorithm estimates the phase shifts needed to implement programming steps as qubit phase changes.

The quantum phase estimation algorithm has several applications in quantum simulation and order-finding. In quantum simulation, the algorithm can be used to simulate quantum systems with a large number of particles. The algorithm can estimate the phase angles of the quantum states that represent the particles in the system. The estimated phase angles can then be used to calculate the properties of the system.

In order-finding, the algorithm can be used to find the order of an element in a finite group. The order of an element is the smallest positive integer that satisfies the equation $e^n = 1$, where e is the element and n is the order.

Applying Simon's Algorithm

Simon's algorithm is a quantum algorithm that solves a classical computing problem with an exponential quantum speedup (the best kind). Daniel Simon, the creator of Simon's algorithm, says that he looked at early quantum computing algorithms and was unimpressed. He was a quantum computing skeptic and did not believe that these algorithms — or, perhaps, any algorithm — would result in a quantum speedup over classical computing.

Simon set out to prove this mathematically, but instead, he created a quantum algorithm that is an important tool and building block in the race to demonstrate broad-based quantum advantage. Simon's algorithm is very valuable for its own sake, and it contributed strongly to the creation, shortly afterward, of Shor's breakthrough algorithm.

Simon's algorithm consists of two main parts: the quantum part and the classical part. The quantum part of the algorithm uses quantum operations to obtain a superposition of all possible values of the period. The classical part of the algorithm uses linear algebra to extract the period from the superposition.

TECHNICAL STUFF

Simon's algorithm is one good example of many processes that need rapid interaction between quantum and classical computers. Supercomputers are being produced today with built-in high-speed interconnects. These allow future quantum computers to plug in and interoperate at very high speeds to solve complex problems, using the best of classical and quantum computers.

Now, let's take a step-by-step breakdown of Simon's algorithm and its applications in cryptanalysis and codebreaking.

SIMON SAYS BEEF UP YOUR SECURITY

Simon's algorithm is a breakthrough in quantum computing that has the potential to revolutionize the field of cryptography and security. It's a powerful tool that enables us to decrypt encrypted messages and crack codes that are otherwise impossible to break using classical computers. Simon's algorithm works by identifying patterns in large amounts of data and using these patterns to find the secret key used to encrypt and decrypt messages. This makes it an essential tool for security experts who must protect sensitive information from prying eyes and hackers.

The problem that Simon's algorithm solves is an interesting problem in computer science and mathematics that requires finding the period of a black box function that is guaranteed to have period 2^n. The *black box function* is a function that takes as input an n-bit string and returns an n-bit string. The period of the black box function is defined as the smallest non-negative integer r such that $f(x) = f(x + r)$ for all x in $\{0, 1\}^n$. The problem is to find the period, r, of the black box function.

The step-by-step breakdown of Simon's algorithm is as follows:

1. Initialize two n-qubit registers to the state $|0\rangle^n$.
2. Apply the Hadamard transform to the first register to obtain the superposition of all possible input strings: $|\psi\rangle = (1/\sqrt{2^n}) * \Sigma x \in \{0, 1\}^n |x\rangle$.
3. Apply the black box function to the first register to obtain the superposition of all possible outputs: $|\varphi\rangle = (1/\sqrt{2^n}) * \Sigma x \in \{0, 1\}^n |f(x)\rangle$.
4. Apply the Hadamard transform to the first register again to obtain the superposition of all possible input strings that have the same output: $|\psi'\rangle = (1/\sqrt{2^n}) * \Sigma x, y \in \{0, 1\}^n (-1)^x \cdot y \cdot f(x) |y\rangle$.
5. Measure the first register to obtain a random string s, which is a linear combination of the possible periods: $s = \Sigma i = 1^r a\_i\ x\_i$, where a_i are random coefficients and x_i are the possible periods.
6. Solve the system of equations to obtain the period *r*.

Simon's algorithm has several applications in cryptanalysis and codebreaking. One of its applications is in breaking the RSA cryptosystem, which is based on the difficulty of factoring large integers.

Simon's algorithm can efficiently factor large integers by reducing the problem to Simon's problem, and then using quantum computing to solve it rapidly.

Another application of Simon's algorithm is in solving the discrete logarithm problem. The discrete logarithm problem is the problem of finding the exponent x in the equation $g^x \equiv h \pmod{p}$, where *g*, *h*, and *p* are known. Simon's algorithm can solve this problem in polynomial time by reducing it to the Simon's problem.

Implementing the Quantum Fourier Transform (QFT) Algorithm

The *Fourier transform* algorithm is a mathematical tool used to transform a time-domain signal into a frequency-domain signal. The Fourier transform algorithm has many applications in signal processing, image analysis, and quantum mechanics.

The *discrete Fourier transform (DFT) algorithm* is a mathematical tool used to transform a sequence of discrete time-domain data into a sequence of complex frequency-domain data. This transformation is achieved by breaking down — decomposing — the original data into its constituent sinusoidal frequencies and representing each frequency as a complex number with a magnitude and phase.

The DFT algorithm is widely used in signal processing, audio and video compression, and image processing. It has applications in many fields, including telecommunications, medical imaging, and geophysics. The algorithm has many variations and optimizations, and its efficient implementation is critical for its widespread use.

Another variant, the fast Fourier transform (FFT) algorithm, shows up in an amazing number of disciplines. Figure 14-5 depicts the decomposition process that the classical FFT algorithm performs for waveforms.

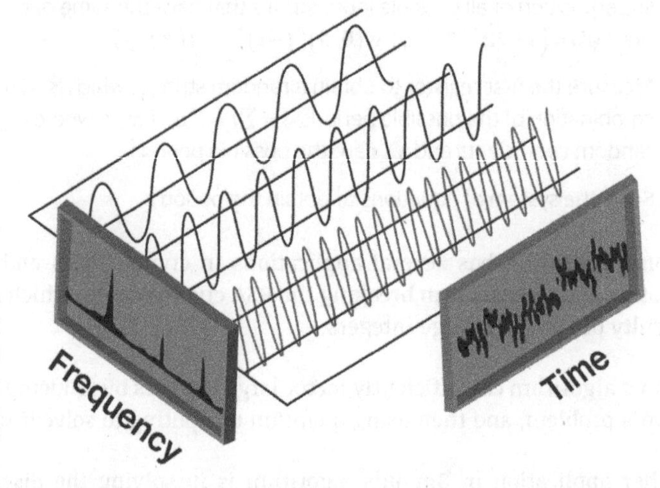

FIGURE 14-5: The classical fast Fourier transform decomposes a signal into its constituent frequencies.

In fact, the FFT reminds us of a joke. Why did Beethoven insist on being buried with a stack of his musical scores — and a pile of erasers? Because he was told he'd be spending many years decomposing.

So that's a lot of background, but the classical Fourier transform algorithms are important, and so is the quantum version. In fact, the quantum Fourier transform algorithm is fundamental to many of the algorithms that convincingly show exponential quantum speedup. It can also be used, directly or with modification, to replace and greatly speed up many applications for which the classical Fourier transform algorithm and its variants are currently used.

One of its most important applications for the quantum Fourier transform algorithm is in quantum simulation, where it's used to simulate the behavior of quantum systems in a controlled environment. Another use for the quantum Fourier transform algorithm is to transform a quantum state into its Fourier series coefficients, which can then be used to calculate the probability distribution of a quantum state. The quantum Fourier transform can also be used in signal processing, including to analyze the frequency content of a signal to identify patterns in the signal, which can be used to make predictions about future behavior.

The comment has been made that most quantum computing algorithms are simply elaborations on QFT. This is an oversimplification, but understanding the FFT algorithm on classical computers and then studying the workings of the QFT on quantum computers is a great starting point for understanding existing quantum computing algorithms — and perhaps for creating your own.

Stepping into Vaidman's Quantum Zeno Effect

The *quantum Zeno effect is* a fascinating concept that has been explored by scientists and researchers for many years. This effect is based on the idea that a quantum system can be prevented from evolving by measuring it frequently enough. Essentially, when a quantum system is being observed, it's prevented from changing. This effect has many interesting applications in quantum measurement and state stabilization.

The quantum Zeno effect is not an algorithm per se at this time; it hasn't been put into the form of code that you can add to a quantum computing program. Instead, it can be used for controlling and measuring qubits.

The quantum Zeno effect was first proposed by the physicist George Sudarshan in the 1970s. He named it after the Greek philosopher Zeno of Elea, who is famous for his paradoxes.

The original Zeno effect has several versions. The simplest one posits that a moving object, such as a running person, approaches a stationary object, such as the end of a race, by cutting the remaining distance by half, cutting that distance by half, and so on. If each step in the process takes some finite amount of time, completing all of them would take an infinite amount of time, so the person can never quite finish their race. The moving object (the system) never quite touches the wall (the solution).

Figure 14-6 illustrates Zeno's paradox. Please don't spend too much time thinking about this, or you might develop Zenophobia (ha ha).

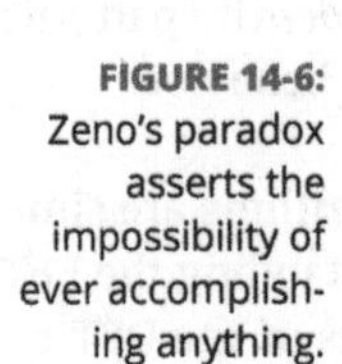

FIGURE 14-6: Zeno's paradox asserts the impossibility of ever accomplishing anything.

One of the most interesting applications of the quantum Zeno effect is in quantum measurement. In quantum mechanics, the act of measurement can change the state of a system, which is known as the *observer effect.* By using the quantum Zeno effect, researchers can prevent the observer effect from occurring, which means that they can make measurements without changing the state of the system they are measuring.

Another interesting application of the quantum Zeno effect is in state stabilization. When a quantum system is in a state of superposition, the state is fragile and easily disrupted by external factors. By using the quantum Zeno effect, researchers can stabilize these states and prevent them from collapsing.

The quantum Zeno effect is a powerful tool that has the potential to advance the field of quantum mechanics.

Getting Linear with the HHL Algorithm

Linear systems of equations are a common problem in many scientific fields. They arise in areas such as physics, engineering, and economics. A system of this type is typically represented by a set of equations that need to be solved simultaneously. The HHL algorithm provides a quantum-enabled solution to this challenge.

The *HHL algorithm,* which was developed by Aram Harrow, Avinatan Hassidim, and Seth Lloyd in 2009, solves linear systems of equations with exponential speedup over classical algorithms.

The *linear systems of equations problem* can be represented as *Ax=b*, where *A* is an *n* x *n* matrix, *x* is a vector of length *n*, and *b* is a vector of length *n*. The goal is to find the values of *x* that satisfy the equation.

Classical algorithms for solving this problem require $O(n^3)$ time, which becomes impractical for large values of n. The HHL algorithm, on the other hand, provides a solution in $O(\log n)$ time.

The HHL algorithm works by encoding the matrix A and vector b into a quantum state. This quantum state is then transformed using a series of quantum operations to obtain a new quantum state that contains the solution to the linear system of equations. The algorithm consists of the following steps:

1. Quantum state preparation: The matrix A and vector b are encoded into a quantum state using a quantum version of the classical Fourier transform.
2. Phase estimation: The phase of the quantum state is estimated using a quantum algorithm called phase estimation.
3. Inverse quantum Fourier transform: The inverse Fourier transform is applied to the quantum state, which results in a new quantum state that contains the solution to the linear system of equations.
4. Measurement: The solution is obtained by measuring the quantum state.

The HHL algorithm has many applications in scientific simulations and optimization. It can be used to solve linear systems of equations in areas such as quantum chemistry, machine learning, and cryptography.

For example, in quantum chemistry, the HHL algorithm can be used to simulate the behavior of molecules, which is important for developing new drugs and materials. In machine learning, the HHL algorithm can solve large-scale optimization problems, such as those that arise in deep learning. In cryptography, the HHL algorithm can be used to break certain cryptographic protocols, which has important implications for security.

Solving and Simulating with QAOA

The quantum approximate optimization (QAOA) algorithm is a hybrid quantum classical algorithm that can solve combinatorial optimization problems. It works by creating a quantum superposition of all possible solutions to a given optimization problem and then measuring the superposition to obtain the optimal solution.

The usefulness of QAOA in quantum computing lies in its capability to solve optimization problems that are difficult or impossible to solve using classical computers. These problems include, but are not limited to, issues in areas such as finance, logistics, and chemistry. QAOA can also be used to simulate quantum systems, which is of great interest to researchers in quantum computing.

One of the pros of QAOA is that it is relatively easy to implement and doesn't require quantum gates, making it a viable option for quantum computing capabilities that are available today and improving steadily, including quantum-inspired computing (Chapter 8) and quantum annealing (Chapter 9). Another advantage of QAOA is that it can be used to solve a wide range of optimization problems, which makes it a versatile algorithm that can be applied to many fields.

TECHNICAL STUFF

Gate-based quantum computers (Chapters 10 and 11) show enormous promise and may someday partly or entirely replace quantum-inspired computing and quantum annealing. However, today and for the near term, they are not yet capable enough to match or beat these less complex quantum computing capabilities.

However, QAOA is not guaranteed to find the optimal solution to a given optimization problem. It also requires many qubits and many gates to solve complex optimization problems, because the number of qubits and gates required increases exponentially with the size of the problem. This requirement means that it's difficult to solve complex optimization problems on today's quantum computers, with their low coherence times.

As a result of this challenge, QAOA will be able to handle larger problems only as the power of quantum computers grows to the needed level. (Don't expect "cheap and cheerful" deals on the exponentially increasing processing costs for problems that exist on the cutting edge, but that's what makes quantum computing in today's environment interesting.)

Getting Grounded with VQE

The variational quantum eigensolver (VQE) algorithm is designed to find the ground-state energy of a given Hamiltonian by using a quantum computer. This is an important problem in quantum chemistry, where the Hamiltonian represents the system's energy. The VQE algorithm can be used to find the minimum point of a molecule, which can be used to determine the molecule's properties.

The VQE algorithm is a hybrid algorithm, requiring a combination of classical and quantum computing capabilities. The classical computer is used to optimize the parameters of the quantum circuit, while the quantum computer is used to evaluate the circuit's energy. This hybrid approach allows the VQE algorithm to be more efficient than other algorithms that use only quantum computing.

However, the VQE algorithm requires many measurements to obtain an accurate result because it uses a variational approach, which requires multiple runs of the quantum circuit with different parameters. Each run of the circuit involves a set of measurements, which can be time consuming and resource intensive.

The VQE algorithm requires multiple runs of a problem using different parameters — and one of the main techniques for error mitigation in quantum computing is to run the same problem many times using the same parameters. You can see the potential for running up a large bill for quantum computing time when rerunning the algorithm many times to account for all of these requirements.

Another disadvantage? The VQE algorithm relies on the capability of the quantum computer to perform accurate measurements and maintain coherence, so any errors or noise in the quantum computer can affect the accuracy of the VQE algorithm. Today's noisy quantum computers are not ideally suited to these challenges.

Assessing Additional Algorithms

Following are additional algorithms that show promise as quantum computers grow in power and accuracy. First, we'll list additional algorithms relating to quantum Fourier analysis, as shown on the left side of Figure 14-1:

- **Quantum counting algorithm:** An offshoot of Shor's factoring algorithm, quantum counting can be used in optimization problems, cryptography, and machine learning.
- **Quantum Fourier sampling algorithm:** This quantum version of the Fourier transform finds the period of a function and is widely used in many algorithms.
- **Quantum phase estimation algorithm:** This algorithm's full name is quantum phase estimation with a quantum neural network (QPE-QNN). Estimating the phase of a quantum state is useful for simulation problems and for developing new and improved quantum computers. The algorithm is promising for use in quantum cryptography as well. When run on a quantum neural network — a machine learning structure with quantum characteristics — large data sets can be processed in parallel. ChatGPT is an example of a powerful application that could become even more powerful using these approaches.
- **Amplitude estimation algorithm:** This algorithm estimates the amplitude of a given quantum state, a crucial step in quantum computing applications such as quantum phase estimation, quantum search algorithms, and quantum simulation. The algorithm uses a combination of quantum operations and classical post-processing to estimate the amplitude accurately and efficiently.
- **Quantum principal component analysis (PCA) algorithm:** Principal component analysis is used in statistics and machine learning as well as image processing and data compression. Quantum versions promise speedups on larger and larger data sets.

To finish listing additional algorithms, we'll mention several in the search and optimization area, as shown on the right side of Figure 14-1:

- **Quantum support vector machine (SVM) algorithm:** The classical support vector machine algorithm is widely used for classification and regression tasks, and the quantum version has specific application to machine learning, including in fields such as finance and healthcare.
- **Quantum walks algorithm:** This algorithm, related to random walk algorithms in classical computing, focuses on a quantum particle moving through a lattice structure. The quantum version is likely to speed up tasks such as search and optimization, as well as helping with quantum simulation.
- **Hidden subgroup problem algorithms:** As the name implies, these algorithms help to find hidden structures or patterns in large data sets. They're useful in computer science, mathematics, and physics. Quantum speedup can make the algorithms far more powerful, yielding speedups useful in cryptography, data analysis, and machine learning.
- **Quantum matrix inversion algorithm:** When run on classical computers, the matrix inversion algorithm is an efficient way to invert large matrices with unmatched efficiency. It's used in optimization problems and machine learning. The quantum version promises to invert matrices in polynomial, rather than exponential, time, promising a larger and larger speedup as the size of the matrix to be inverted grows.
- **Quantum k-means algorithm:** The classical k-means algorithm is used for clustering. The classical version scales with the number of data points, whereas the quantum version scales with the square root of that number — a truly significant speedup.
- **Quantum approximation of nonlinear functions (QANF) algorithm:** This quantum approach will bring speedups to algorithms used in finance, engineering, medicine, and elsewhere.

Identifying What's Ahead

Well, that was a lot to grasp, wasn't it?

If your brain doesn't hurt at this point, we, the authors, have either done an absolutely astonishing job of explaining quantum algorithms — or we failed miserably. Seriously, it's okay if your head hurts; we know ours did writing this chapter.

Each of these quantum algorithms has the potential to solve problems in seconds, minutes, or hours that currently might take years, centuries, or millennia to compute classically. This speedup will happen for more and more challenging problems as quantum computers mature.

At the forefront of this research are a number of universities and organizations. For example, IBM is working on developing a suite of quantum algorithms that can be used for a variety of applications. Meanwhile, researchers at the University of California, Santa Barbara are exploring the use of quantum algorithms for optimizing complex systems.

Another organization doing important work in this area is the Canadian Institute for Advanced Research. They have a team of researchers dedicated to developing quantum algorithms that can be used to solve problems in fields as diverse as finance, physics, and biology.

A lot of hard work is ahead on all fronts before the potential of quantum computing can be more fully realized. An important part of that work will be the refinement of existing quantum computing algorithms and the development of new ones. But today's quantum computing algorithms are a strong starting point and promise to serve as a kind of mathematical ski jump, launching the discipline of quantum computing well clear of previous technologies. We hope we've given you a feel for what's possible, today and in the future.

» Examining quantum computing vendors

» Seeing what Strangeworks has to say

Chapter 15
Cloud Access Options

Getting started in quantum computing is greatly helped by the advent of the cloud. You can easily get access to quantum computing simulators and smaller systems today. In addition, plenty of courses (see Chapter 16) use available cloud resources to help you get started. Free credits are even available for many systems, reducing the cost of learning.

We're also happy to put forward this book as an important resource. In these pages, we have, we hope, helped you get familiar with the major concepts and the opportunities available to those who learn how to do quantum computing. This should save you time and frustration as you get started.

Cloud access has received a lot of credit — deservedly so, we believe — for opening up access to quantum computing. As a result, quantum computing may be the most democratic advancement in high-performance computing since the invention of the abacus.

In this chapter, we describe the major cloud interfaces for quantum computing access, including programming options. These descriptions will not only help you get started; they'll guide you in looking at alternatives if your first choice runs out of gas in some way.

CLOUDY AT THE SUMMIT

If you're doing intermediate to advanced work in quantum computing, you'll face two barriers: expense and access.

If you start to burn serious cycles on any kind of quantum system — quantum-inspired computing systems (Chapter 9), quantum annealers (Chapter 10), or gate-based quantum computers (Chapters 11 and 12) — costs can go up quickly. And it's such a new field that you can't be sure how cost effective your efforts will be until you try.

At the most advanced levels, even if you get the budgetary resources, availability becomes an issue. There are only a few of the highest-end quantum computers around, and time on them is sliced thinly indeed. One researcher described waiting in an online queue for nearly 24 hours, just to gain 10 seconds of processing time on one of the most advanced quantum computers in existence today. The work got done, but the process was frustrating.

Use the resources here to learn and make connections so you can make smart choices as your work in quantum computing advances.

Exploring the Major Types of Options

There are three kinds of options for cloud access to quantum computers: public cloud providers; device manufacturers; and Strangeworks and a few other similar access providers. We describe each option briefly here, and in more detail in the pages that follow.

Grouping public cloud providers

The three public cloud providers — AWS, Microsoft Azure, and Google Cloud Platform — offer access to quantum computing options. These options are a great way to get started if you're already a customer of one or more of these providers.

However, each provider has their own partner relationships and business interests. None of the public cloud providers offers all possible options, and each is, known to be or said to be developing its own quantum computers. So they are both cooperating with, and competing with, established quantum computer providers.

This "coopetition" means that each provider can offer plenty of help to get you started, but it may slow down their capability to get access to the latest and greatest for their customers. It may also influence the quality of advice and access you get for each platform. And the public cloud providers may or may not give you a way to escalate an issue that you have with a given provider's system to that provider's support staff. For instance, if you have a problem with a quantum computer from provider X, you may not be given a way to reach provider X's support staff through the platform.

Similarly to the public cloud providers, quantum computing is just one offering in the overall world of high-performance computing, or HPC. The public cloud providers are interested in selling computation time — they call it *compute* — and storage space. They aren't too bothered as to whether it's quantum compute or not; they will put their resources where the money is at this moment. So if you believe that quantum computing has a huge upside, and you have a particular interest in getting going with quantum computing over more established technologies, you may find support a bit thin as you go forward.

The public cloud providers can also make it almost too easy to get started. One, or several, developers from your company may get started with quantum computing at a low cost, but as they keep going, things get more expensive. It reminds us of a joke that applies to the entire field of cloud computing: "There's just one problem with 'pay as you go': As you go, you pay."

Table 15-1 summarizes the quantum computers available on major public cloud providers at this writing. For comparison, we also provide the list for a major independent access provider, Strangeworks. Each of these offerings is described in more detail in this chapter.

Examining quantum computer manufacturers

Most of the major quantum computer manufacturers provide cloud access to their systems. This is a wonderful option to have compared to before the advent of the cloud, when you had an all-or-nothing choice with computers: You could either do without or buy an entire system. There was no easy way to pay for variable access to a system, whether that was on a trial basis or for a system you wanted to make some commitment to.

With the cloud, quantum computer manufacturers can offer you cloud access on a fractional basis — as little or as much as you need.

TABLE 15-1 **Public Cloud Providers and Quantum Computer Manufacturers**

Company	Technology	Amazon Braket	Azure Quantum	Google Quantum AI	Access Providers*
D-Wave	Quantum annealing				x
Google	Superconducting			x**	
IBM	Superconducting				x
IonQ	Trapped ion	x	x	x	x
OQC	Superconducting	x			x
Pasqal	Neutral atom		x		x
Quantum Circuits, Inc.	Superconducting		x		x
QuEra	Cold atoms	x			x
Quantinuum	Trapped ion		x		x
Rigetti	Superconducting	x	x		x
Xanadu	Photonic				x
Total		4	5	2	10

**This specific list of quantum computers is for Strangeworks; others have somewhat different lineups.*

***Google's quantum computers are available only to selected applicants using the Google Quantum AI platform.*

But before you commit resources to a specific manufacturer, you should learn as much as you can about the manufacturer and what they offer. Among the factors to consider are the following:

- Maturity of quantum hardware
- Longevity and competitive position of the underlying qubit technology
- Capability to develop all the other parts of a quantum computer
- Transparence about the likely cost of using the system
- Positive results or complaints from previous customers
- Customer support
- Access models — cloud portals and so on
- Contractual and IP requirements

NOW YOU SE ME, NOW YOU DON'T

One of the most crucial roles in computing today is the role of the sales engineer or solutions engineer (SE for short, either way), sometimes called a solutions architect (SA) or similar titles.

Solutions engineers most often work as part of a company's sales team and are paid a high salary, plus a commission determined by the company's sales success.

SEs are valuable because they serve as an interface between the customer's technical issues with a hardware or software system and the vendor's engineering team. The job of the SE is to help troubleshoot and to provide the "glue" between the customer's systems and approach and the specific products offered by their employers.

As a customer, the quality and availability of SEs for a given platform is one of the most important factors in determining how likely you are to be able to use that platform effectively. This is especially true in a new and technically challenging area such as quantum computing.

Before you make a commitment to a given access method for quantum computing, do what you can to find out how strong the SE or similar support resources are, and how much they're tied to your spending level on a platform. Whatever you learn, take it into account as you decide how best to move forward. Getting solid SE support may make a big difference in your ability to make progress.

It's a big commitment for you and your company to choose a provider. You'll spend time and money just learning your way around and finding out whether a specific device maker offers something that's worth your focus over time.

Several device manufacturers have done a good job of making it easy to get started with cloud access to their specific systems, but you may find this almost too easy. You could get a long way down the road with one manufacturer before you start to wonder if a competitor, or a platform that offers choices, may have been a better bet.

By then, however, you will have made a big investment in technical work and relationship-building with a given manufacturer and their cloud platform. You may find yourself with the problem so eloquently stated by the rock group The Eagles in their famous song, *Hotel California:* "You can check out any time you like, but you can never leave."

All the time and money that you spend on one provider's offerings is not, by definition, available to spend on one of the others. So take your time experimenting with different options before you commit to any one device maker's cloud access platform.

WARNING

Be careful of costs when using quantum computing offerings. It's a new field, which makes it easy to get surprised by cost overruns. Many platforms offer cost ceilings that halt a job once the ceiling is reached. Use these ceilings where they're offered; they can save you a lot of unnecessary expense. You don't want the most dramatic entanglement you experience in quantum computing to be entanglement between a quantum computing vendor and your bank account.

Another issue is that offering a solid public access offering is just one part of the distribution strategy for each provider. The best access may not mean the most appropriate provider for your needs. And different vendors may make it easier or harder to gradually scale your involvement to meet your needs, rather than theirs. They may also vary in their commitment to important issues such as security, uptime, and access to the latest and greatest hardware.

The final issue is that the entire field is still fluid. A given manufacturer is going to have expertise on only their own systems. When you have a problem to solve, they will not be able (or motivated) to help you look at the big picture of solutions, if those approaches are from a different maker or use a different technology.

Looking at access providers

First, a note of disclosure: Strangeworks, a leading access provider, was founded in 2018 by one of the authors (whurley); he is CEO of the company today. So we, as an authoring team, may not be able to be completely objective in assessing Strangeworks as an option. But we do promise to try.

Several companies provide quantum computer access as a core part of their business. This access functions somewhat like the quantum computing offerings on AWS and Azure, but usually with a wider range of systems supported.

Leading access providers include the following half-dozen companies:

- **Agnostiq** (`https://agnostiq.ai`) was founded in 2018 and is based in Toronto. They are shown on the startup website Crunchbase as having received a Series A round of funding and $28 million in total funding.
- **Classiq** (`www.classiq.io`) was founded in 2020 and is based in Tel Aviv. Crunchbase shows them as having reached Series B funding, with $61.8 million in total funding.

- **QBraid** (www.qbraid.com) was founded in 2020 and is based in Chicago. According to Crunchbase, they are privately held and not VC-funded, so an accurate funding level is not available.
- **QCWare** (www.qcware.com) was founded in 2014 and is based in Palo Alto, California. They are privately held, but Crunchbase shows them with $41.4 million in total funding.
- **Strangeworks** (https://strangeworks.com) was founded in 2018 and is based in Austin, Texas. They have reached the Series A level for funding and have received $28 million in total funding.
- **Zapata** (www.zapatacomputing.com) was founded in 2017 and is based in Boston. They have reached the Series B funding level and have received $67 million in total.

Other, similar companies may be less well known or less well funded. Also, different companies may tend to support specific geographical areas, in terms of their hours for technical support. Check with each company you're interested in before choosing an access provider.

Each of these companies provides access to multiple resources, which in most cases include the cloud provider offerings Amazon Braket and Azure Quantum and quantum computing systems from D-Wave, IBM, IonQ, and others.

In addition to offering access, each access provider is likely to offer various kinds of customer support, a platform for access, educational offerings, and community. Getting access to the specifics of each access provider's offering may require being a customer, so we have not provided a detailed description for each provider. Instead, we describe Strangeworks as an example, later in this chapter.

Noting the Importance of Amazon Braket

The Amazon Braket quantum computing service from AWS is cleverly named after the bra-ket notation used to describe vectors with complex elements, as needed in quantum computing, including computing. The service, introduced way back in 2019, includes the Braket Development Environment, which uses Python and an interface organized around the popular Jupyter notebook approach. It's an outgrowth of the AWS Center for Quantum Computing, which fosters collaboration between researchers, academics, and industry experts to advance quantum computing hardware and software development.

At this writing, Braket offers quantum computing hardware from IonQ, OQC, QuEra and Rigetti. IonQ uses trapped ion qubits; OQC and Rigetti run superconducting qubits; and QuEra uses cold atoms (all are mentioned in Chapters 10 and 11). This flexibility can also lead to difficulties; as a unified platform that runs on multiple providers' systems, Braket may work more effectively on some systems than others. You may find yourself having to dive into the specifics of different systems to solve problems that arise. Cost is also an issue; as you get more serious with your work, you can run up big bills, especially because pricing and performance are not consistent across systems.

As the first, and still the largest, cloud computing service, Amazon has immense loyalty among many of its customers. It also has large, prepaid contracts with many of them, which means that your company may have AWS credits that you can use to learn quantum computing on Braket without incurring new costs for your company.

Braket is the leading cloud provider offering for cloud computing and is part of a wide range of HPC offerings on AWS. The features of Braket include

- **Free tier:** Braket has a free tier to get you hooked — we mean, started.
- **Braket SDK:** This software development kit (SDK) allows you to run quantum algorithms on different quantum computing hardware and simulators, each of which has its own cost. (Since execution times vary, you have to experiment to see what the total cost is to execute a given job across different quantum computing platforms.)
- **Pulse-level access:** This low level of control over applicable quantum computing hardware allows you to manage qubits directly. However, this feature is available for only some hardware. Also, if you're doing a lot of work at this level, you're more likely to get the highest level of control and support by working directly with a manufacturer.
- **PennyLane integration:** The open-source software library PennyLane is integrated with Braket. PennyLane is a quantum computing software library that is machine learning–oriented, supported by quantum computing vendor Xanadu and with the capability to interface to machine learning resources PyTorch and TensorFlow.
- **OpenQASM support:** OpenQASM is an open-source intermediate representation for gate-based quantum computing programs. OpenQASM seeks to run at a low level, close to the hardware, without being tied to any one hardware provider.
- **Simulator support:** You can run quantum computing programs on simulators, which run on classical computing hardware, at a lower cost than quantum computing hardware. This can save you money because you can

improve and optimize your software less expensively on classical computing hardware before trying it on quantum computing hardware.

- **Analog Hamiltonian Simulations support:** Separately to gate-based quantum computing, Hamiltonians represent a specific set of simulation problems. (The popular musical *Hamilton!* can be considered a simulation, and it is Hamiltonian, but it's not *a* Hamiltonian.)
- **Hybrid jobs support:** Braket has support for jobs that span classical and quantum computing. This is certainly convenient, but the pricing is hybrid too, so be careful about the size of the bill you're running up as you use hybrid programs.
- **Quantum Solutions Lab availability:** You can engage with the Amazon Quantum Solutions Lab, which contracts out for SE-type support for your work in Braket.
- **Standardized pricing and billing:** AWS has done a nice job of standardizing pricing and billing across providers, making it easier to manage your spending as you make increasing use of quantum computing.

TECHNICAL STUFF

The Braket SDK lets you run the same program on any of several providers, but the underlying systems differ considerably. As a result, the Braket SDK may support some providers less well than others. Also, AWS does not tend to connect Braket customers with quantum computer providers for optimizing customers' software nor other in-depth support.

Figure 15-1 shows the console for Amazon Braket, allowing you to manage quantum computing jobs.

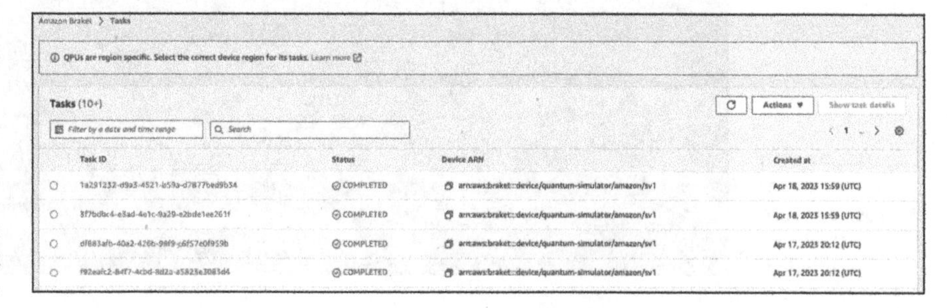

FIGURE 15-1: Amazon Braket console.

To summarize a very complex offering: If you're a loyal AWS user with lots of credits handy, there's no reason not to get started in this robust offering, especially if you can stay in the free tier, or at least keep your costs down while you're on the early part of your learning curve.

After you get some experience under your belt, however, and know what problems you want to focus on, carefully compare offerings from the major cloud providers, quantum computer providers, and access providers such as Strangeworks before proceeding. You'll want to be sure you're getting the most bang for your buck, as the number of bucks you spend can add up quickly.

Counting on Azure Quantum

Microsoft Azure followed AWS to market, but Azure was the first public cloud provider to announce a quantum computing offering, a few years later. However, Azure Quantum was not opened up to broad public use until after Amazon Braket had been announced and launched.

Today, as with Azure itself, Azure Quantum is a wide-ranging offering that can meet most needs. Microsoft and Azure both have a lot of longstanding, loyal customers. You can get off to a good start with Azure Quantum. Figure 15-2 shows a summary of the Azure Quantum offering.

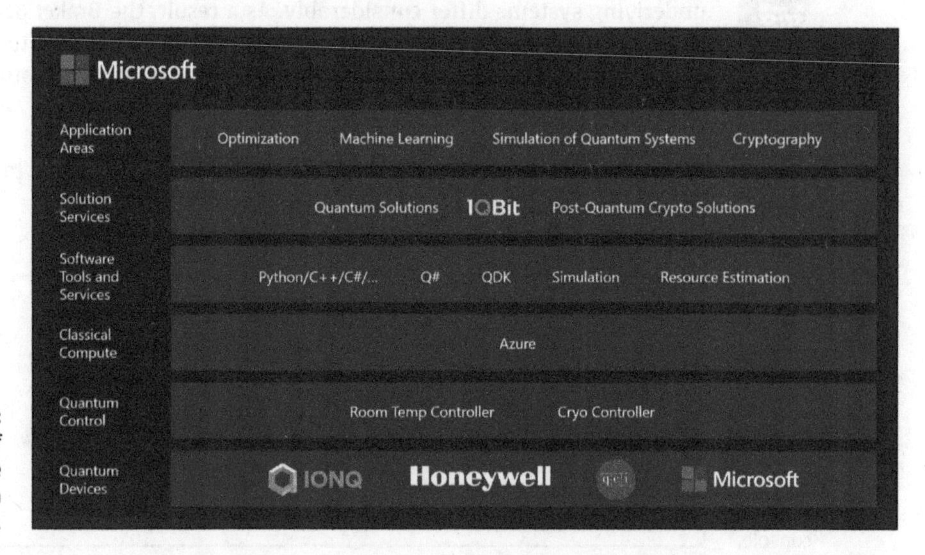

FIGURE 15-2: A summary of Microsoft's Azure Quantum offering.

Azure Quantum, at this writing, offers five providers of quantum computing hardware: IonQ, Pasqal, Quantinuum, Quantum Circuits Inc., and Rigetti. IonQ and Quantinuum use trapped ion qubits, Pascal uses neutral atoms, and Quantum Circuits Inc. and Rigetti use superconducting qubits. Azure also lists several optimization solutions as providers, including 1Qbit algorithms and Toshiba's Simulated Quantum Bifurcation Machine (SQBM), powered by classical computing GPUs.

Although Azure Quantum does not have a defined free tier, unlike Amazon Braket, they currently offer a generous $500 in Azure Quantum Credits for use with each new quantum hardware provider. You're even invited to ask for more if you run out.

The Azure offering is similar to Amazon Braket in that it uses Jupyter Notebooks, offers access to quantum libraries, and offers simulators. You can use the Cirq, Qiskit, and Q# SDKs for programming. It doesn't promote lower-level hardware access nor support services in the same way that Amazon Braket does. It does provide a resource estimator to help you predict job costs.

Figure 15-3 shows a sample notebook populated to run a quantum computing job.

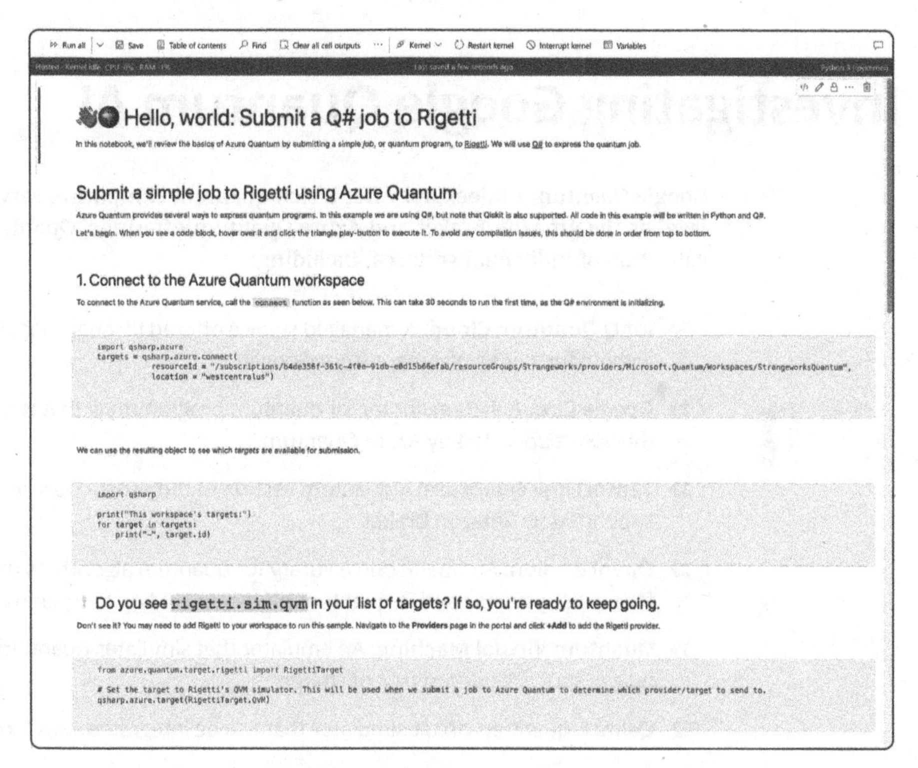

FIGURE 15-3: Using a notebook to run a job in Microsoft's Azure Quantum service.

Microsoft offers Q#, a high-level programming language tailored for quantum computing. It runs on simulators as well as real quantum computers. Q# is integrated with Microsoft's widely used Visual Studio integrated development environment (IDE).

Microsoft is working on quantum computing hardware, including controller chips for quantum computers that operate at room temperature and, separately, at cryogenic temperatures. However, the company has been pursuing a quantum computing strategy based on topological qubits (see Chapter 12). There have been some announcements from this effort, but not much solid progress to date.

You can get off to a solid start with the Azure Quantum offering, especially for optimization problems. Microsoft customers tend to be loyal, and you may find the environment comfortable. And it's a good idea to use free credits to experiment and get an idea of your needs. However, once you get beyond free credits, you should take a careful look around. Look at other cloud providers, quantum computer providers, and providers such as Strangeworks, choosing the offering that gives you the most support for the problems you want to tackle.

Investigating Google Quantum AI

Google Quantum AI does not offer a rich quantum computing service at the same level as the Amazon Braket and Azure Quantum offerings. Quantum AI is more a collection of individual services, including:

- **IonQ Quantum Cloud:** A managed service offered through Google Cloud, without further involvement from Google.
- **Google Cirq:** A Python library for quantum programming that is also among the SDKs supported by Azure Quantum.
- **TensorFlow Quantum:** A quantum version of the TensorFlow library, also supported by Amazon Braket.
- **OpenFermion:** An open-source library for quantum algorithms that simulate fermionic systems used in quantum physics, including for quantum chemistry.
- **Quantum Virtual Machine:** An emulator that simulates quantum-processor-type output; it's offered free of charge.
- **Qsim:** A quantum circuit simulator that can be integrated with Cirq.
- **Access to Google quantum computers:** Access only by contacting Google, not by signing up online.
- **Tutorials:** Educational resources for students and others looking to build quantum algorithms.

Figure 15-4 shows information about Google's Cirq offering.

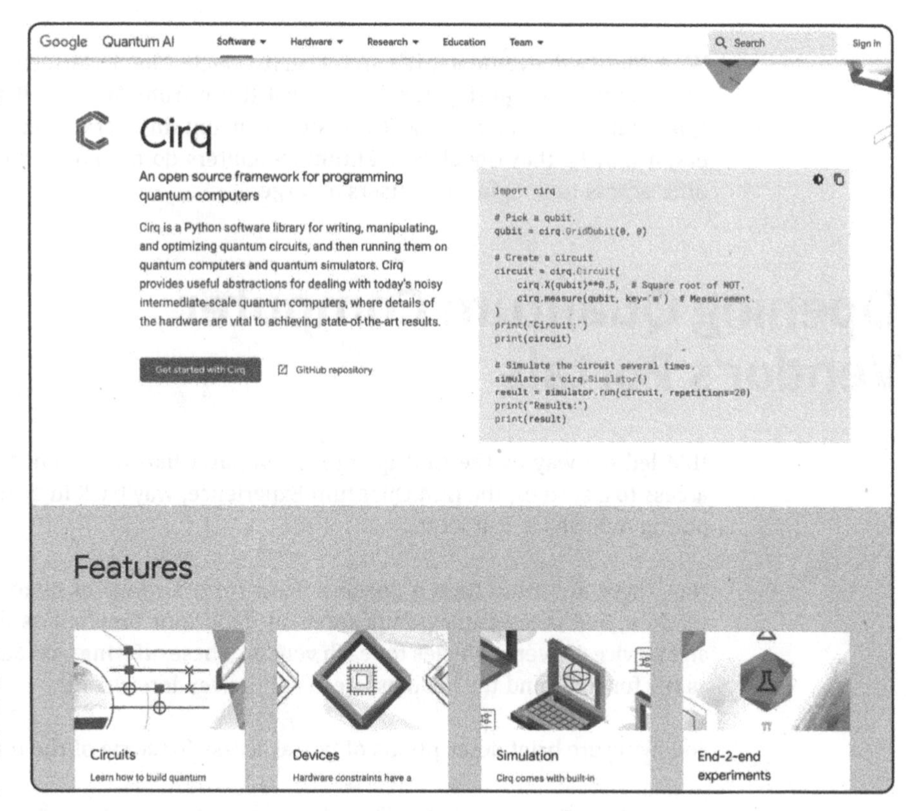

FIGURE 15-4: Google has developed Cirq, a Python software library for quantum circuits.

Cirq, developed by Google's Quantum AI team, was released in 2018. It is now an open-source project. It offers intuitive APIs and comprehensive documentation. Cirq provides various tools and libraries that assist developers in working with quantum circuits, simulating quantum systems, and conducting experiments on real quantum hardware.

Cirq offers extensive customization options. But, as with any quantum computing development, running quantum simulations or experiments can be time consuming and resource intensive.

Google is a leader in two areas vital to quantum computing:

- **Quantum computing hardware:** Google has at times rivaled IBM in quantum computing hardware, using similar superconducting qubits and developing systems with scores of them. Google made the first quantum supremacy claim back in 2019 — a claim quickly disputed by IBM (see Chapter 6).
- **AI and machine learning:** Google Cloud Platform has a solid claim at the leading public cloud provider for machine learning and AI, and has developed the Cirq Python library for quantum circuits.

With these achievements, it's ironic that Google Cloud Platform doesn't offer a full quantum computing service to rival those from Amazon Braket and Azure Quantum. While Google seeks to create and maintain relationships with researchers, it may be that Google's quantum computers do not have enough capacity to offer access to Google Cloud users at large.

Opening Quantum Computer Vendor's Portals

IBM led the way as the first quantum computer hardware vendor to offer cloud access to a system, the IBM Quantum Experience, way back in 2016. Rigetti Computing followed a year later.

Now there are more than a dozen portals from a range of quantum computing vendors, and these entry points serve an important function as marketing tools and service delivery vehicles for each vendor. These offerings are subject to change as the fortunes and the business needs of the vendors change.

Following are brief descriptions of portal access for some of the major vendors:

- **D-Wave The Leap:** D-Wave is no longer available on Amazon Braket, but you can still access D-Wave's quantum annealing systems via The Leap, the company's cloud service. Usage of The Leap includes access to recently shipped systems and quantum hybrid problem-solving capability.
- **IBM Q Experience:** The IBM Q Experience includes some of IBM's most powerful quantum computers. It has its own graphical user interface, Quantum Composer, which emits Python code (see Figure 15-5). It uses the OpenQASM standard (also in Amazon Braket) and includes the Qiskit SDK, a tutorial, and an online community.
- **Nvidia DGX Quantum:** Nvidia is red-hot at this time, with its GPUs offering sorely needed processing capability for AI and machine learning. The same hardware is being used for quantum-inspired computing (see Chapter 9) running on classical computing hardware (Nvidia's GPUs).
- **Quantinuum Quantum Cloud Services:** Quantinuum offers trapped ion quantum computing through its cloud services, with an AI platform and TKET, an SDK accessible through the extensible PyTKET Python SDK.

» **Rigetti Quantum Cloud Services:** Rigetti uses superconducting qubits and hosts Quantum Cloud Services (QCS), which includes their Forest SDK. QCS includes access to an optimizing compiler, Quil, and QVM, Rigetti's own free simulator.

» **Xanadu Quantum Cloud:** Xanadu is a leader in photonic quantum computing. They offer a Python library and a set of applications in a variety of fields, including quantum machine learning, and they maintain the PennyLane open-source software library.

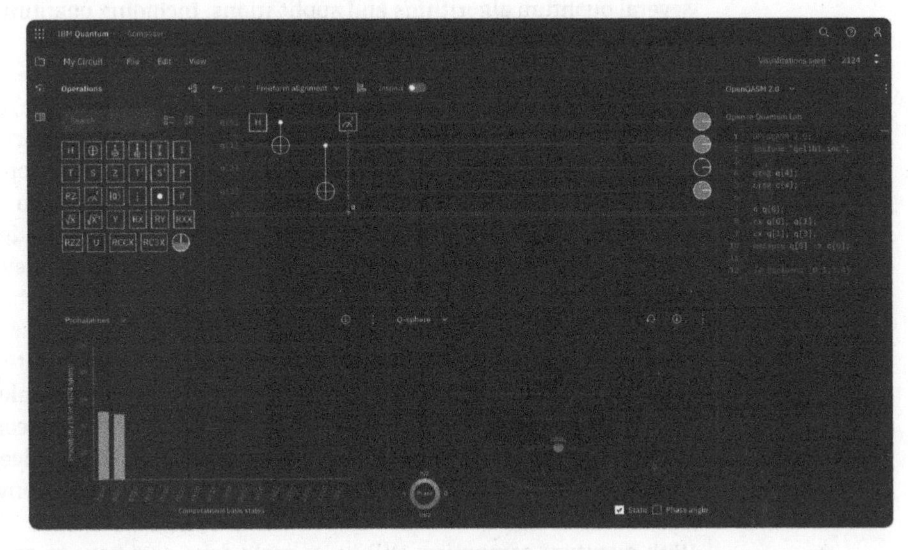

FIGURE 15-5: The IBM Quantum Composer is a no-code processing tool that emits Python code.

D-Wave offers Ocean, a Python-based platform that works with its quantum annealing systems (see Chapter 9). NASA, Google, and Volkswagen are among the many members of the Ocean community. You'll see Ocean users on social media platforms such as Reddit and Twitter. D-Wave offers a relatively easy way in but is quite different from the other platforms listed here, which mostly work with gate-based systems (see Chapters 10 and 11).

IBM's Qiskit is one of the most popular frameworks in the field. Qiskit, an open-source project that uses Python, was first introduced in 2017 by IBM's Quantum team. Qiskit provides tools and libraries that make it easy to work with quantum circuits, simulate quantum systems, and run experiments on real quantum hardware.

Nvidia has recently announced the CUDA Quantum Software Platform. It allows Nvidia's community of high-performance computing users to extend their work into the quantum computing realm and gives new users a rich mix of wholly classical, quantum-inspired, and fully quantum approaches.

Quantinuum's TKET quantum framework, formally known as the Honeywell Quantum Education Toolkit, also supports multiple quantum computing technologies, including Honeywell's trapped ion quantum hardware. TKET offers visualization tools and optimization capabilities, such as circuit rewriting and synthesis algorithms. As a newer platform, it may not yet have the robust community support of some others.

Rigetti offers PyQuil, which is specific to Rigetti's quantum computers. Like other offerings, it's an open-source language based on Python. It's relatively approachable. Rigetti works closely with users, and PyQuil has already been used to develop several quantum algorithms and applications, including quantum chemistry simulations and machine learning algorithms.

Xanadu offers a relatively new quantum framework, PennyLane, designed for ease of use. PennyLane works with a wide range of quantum hardware platforms, including both gate-based and continuous-variable quantum computers. The goal is for developers to write their algorithms once and then run them on various quantum hardware platforms without worrying about the underlying hardware details. With the tremendous variation among platforms, however, your mileage may vary.

TECHNICAL STUFF

Mitiq, which you can find on Github, is a different kind of quantum programming tool. It's a quantum error mitigation library that promises to make quantum computing more reliable and accurate. Tools include error-correction algorithms, noise modeling, and quantum state tomography. If you go deep into quantum computer software and solutions development, Mitiq may become a good friend.

With quantum computing still in its early days, you have to make some choices that don't offer easy answers:

- Some frameworks are optimized for a single vendor's systems, which locks you in somewhat but gives you direct line of sight from coding to execution.
- Other frameworks are multiplatform, which gives you flexibility but requires you to roll up your sleeves and dive in to solve problems or optimize performance whenever needed on each system on which you run the multiplatform framework. Costs may also vary significantly and surprisingly across platforms.
- Some platforms offer simulator access, which gives you a low-cost option for early work, postponing the expense of running on actual quantum computing hardware until you're ready.
- And some platforms offer access to the hot new thing while others are of relatively long standing, with more resources and depth of experience.

» Cost can rear its ugly head in unexpected ways. The cost of trying stuff can be low, and it may even be covered by free credits. But costs may rise quickly just as you start to sink your teeth into solving real problems.

Did we mention yet that there are no easy answers? There is, however, a rich array of options. The rewards for making smart choices and making a contribution to the community during these early days may be significant.

Unlocking Quantum Potential with Strangeworks

TIP

We suggest that you read this section, even if you don't have immediate plans to access Strangeworks. The Strangeworks portal is intended to be educational, so reading about it will give you a good idea as to what you're getting into when you set out to do some real quantum computing.

Strangeworks is designed to bring quantum computing to the masses — yes, that means anyone who has an interest, regardless of how they're coming into quantum computing. The Strangeworks portal provides a unified, user-friendly interface that simplifies the complexities of quantum computing, allowing developers, researchers, and enthusiasts to collaborate, innovate, and build quantum algorithms and applications. You can do initial work in a free demo environment, and then sign up to dive deeper into resources that have a cost associated with them. (Sometimes free credits are available to offset part or all of the cost.)

In addition to compute providers, the platform includes access to Qiskit, the Strangeworks SDK, the Strangeworks Python SDK, and the QAOA algorithm for QUBO problems. There's a community library and a rich community where the entire range of technologies accessible through Strangeworks can be discussed and compared.

Strangeworks serves as a lab bench for the widest available range of systems, a support service for those looking to get started with quantum computing, an accelerator for those who are serious about it, and a warm introduction in case you want to commit serious resources in partnership with one or more providers. It's the broadest of the available offerings.

Strangeworks is a managed service and includes an ecosystem that integrates with existing high-performance computing infrastructure — not only quantum computers — and orchestration platforms used by its clients, including (but not limited to) the Amazon Braket and Azure Quantum offerings. For example, clients can access quantum computing systems through Strangeworks by using Mathematica, MATLAB, or API calls as an interface. The platform provides tools and resources that enable users to understand, design, simulate, and eventually deploy quantum algorithms in real-world applications.

Using Strangeworks begins with a guided tour that starts at the main dashboard, as shown in Figure 15-6. There is a no-cost experience and a premium offering.

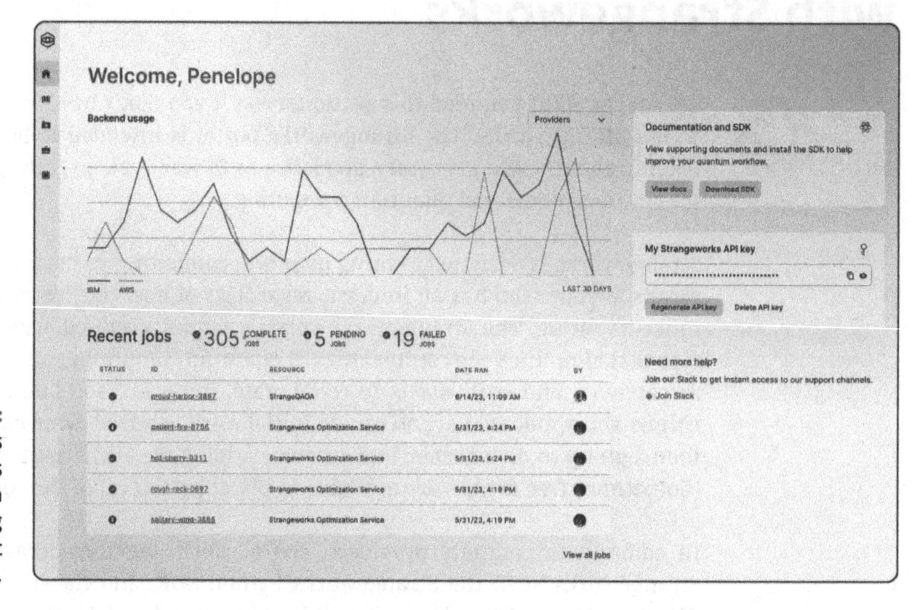

FIGURE 15-6: The Strangeworks portal delivers quantum computing information at your fingertips.

The dashboard provides information on the jobs that have run to date, links to product documentation, and a link to download the Strangeworks software developer kit (SDK), which you use to submit jobs to the platform. It also includes your Strangeworks API key, the gateway into all the technologies in its expansive ecosystem. (For support, the company has an open Slack channel for clients and users.)

TO EXPLORE STRANGE NEW WORLDS

A hidden asset for each portal you might want to use is the team behind it. Like the local fire department, you don't get to know how good their services are until you need them. At that point, you hope to heck they're solid professionals, lest your house — or, in this case, your quantum computing budget — burns to the ground.

The Strangeworks team has expertise in software engineering. They are highly knowledgeable about advanced computing in general and quantum computing in particular. They understand the breakout nature of these cutting-edge technologies, which makes it in their interest to go the extra mile in supporting development on these advanced platforms — not just at the size that the industry is today but with a view toward what it's likely to become.

One of the authors (whurley) plays a key role here. He whurls all over the, well, whurld, giving talks, attending ribbon-quttings (ha ha) for new quantum computers, and more. whurley keeps the Strangeworks name in front of hardware and software providers, as well as users and their organizations. All this activity helps ensure a constant flow of new people and new ideas to the Strangeworks platform, keeping the much-put-upon Strangeworks team — as well as any coauthors he drags into the madness (ahem) — constantly involved with the latest and greatest.

Perusing a classical catalog of quantum technologies

When you use the Strangeworks platform, everything you do is linked back to products in the Strangeworks product catalog, as shown in Figure 15-7.

The list of resources in the product catalog is customized to your specific workspace. For instance, the workspace shown in Figure 15-7 is a demo environment.

You can also scroll down to a list of computer providers, as shown in Figure 15-8. The portal includes a list of every compute provider that Strangeworks makes available — from quantum-inspired computing (that is, running on classical computers) to gate-based quantum computing.

Strangeworks offers a wide range of providers, making it easy to keep up with the latest and greatest in quantum computing(and also giving you the ability to gossip about who's in and who's out of the list over time).

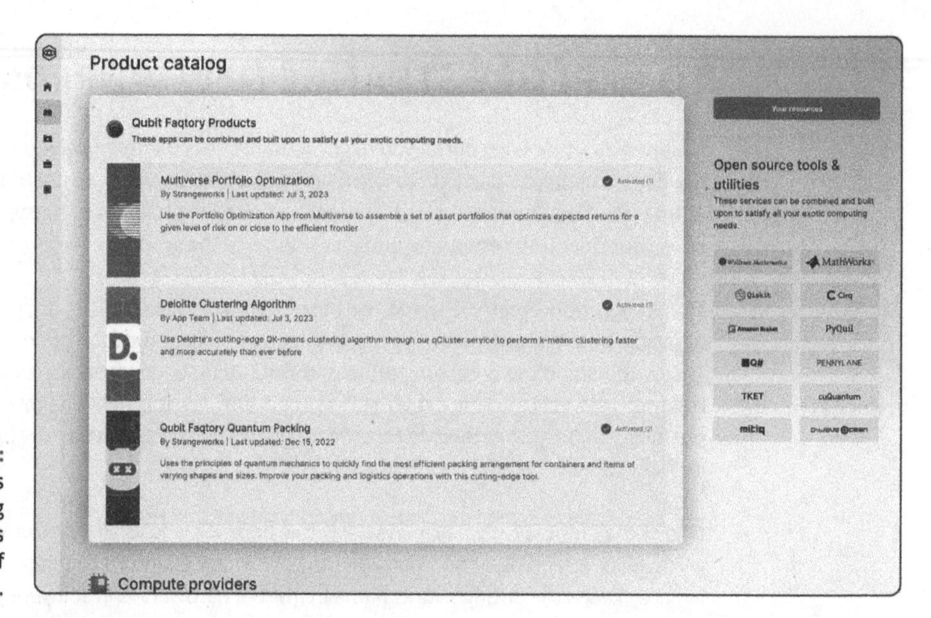

FIGURE 15-7: The Strangeworks product catalog gives you access to a wide range of resources.

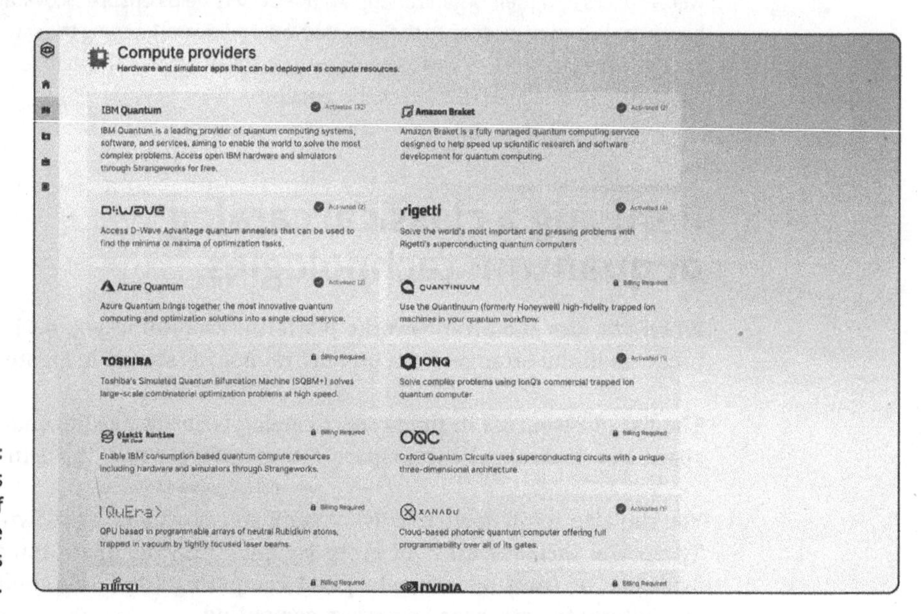

FIGURE 15-8: The Strangeworks "long list" of compute providers is comprehensive.

Exploring compute providers

Each compute provider has its own page with information about what technologies are available, documentation on how to get started, pricing information, and Strangeworks products and services that work with that provider. It's intended to give you easy access, as well as a feel for what's possible. Figure 15-9 shows the compute provider page for IBM Quantum.

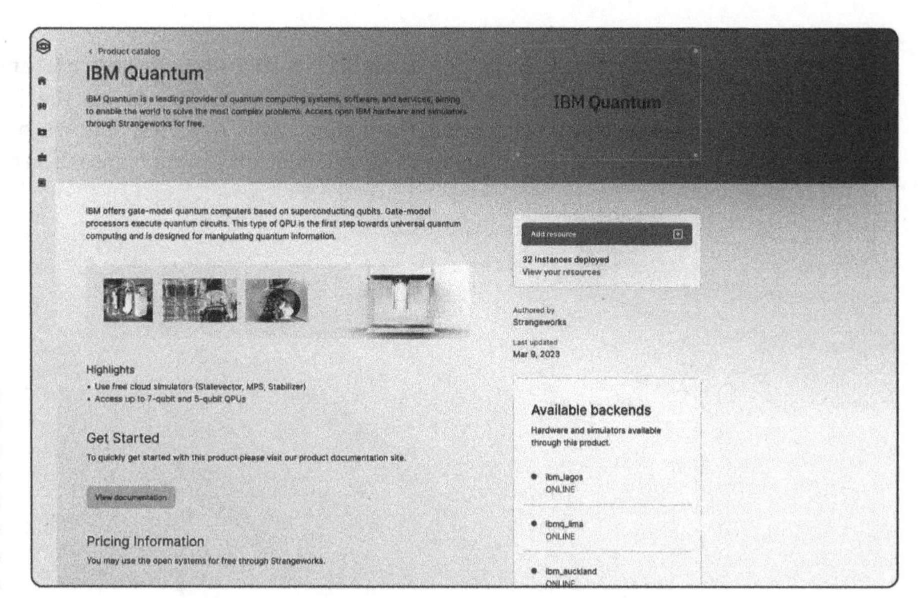

FIGURE 15-9: IBM Quantum is one of the compute providers available in the Strangeworks portal.

To see a summary of terms and conditions associated with a particular product, click Add Resource (on the right of the screen). Then simply accept the terms to activate it.

After you activate a resource, you deploy it to a chosen environment. The resource you've deployed will then appear in — wait for it — a list of deployed resources, as shown in Figure 15-10. You can also change each resource name to fit your needs, such as naming the team using the resource.

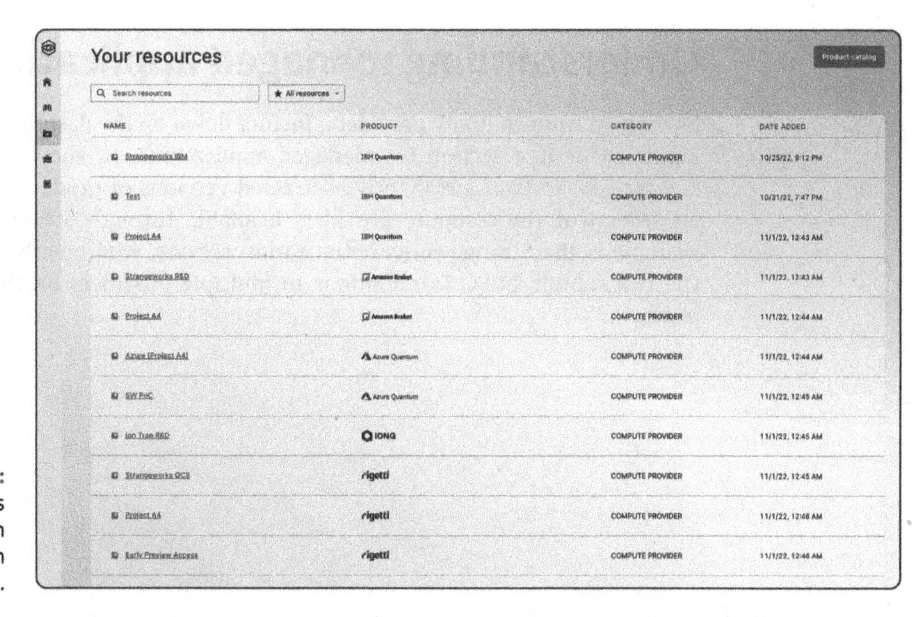

FIGURE 15-10: You can see jobs waiting to run on your chosen resource.

Clicking a specific resource displays its ID. Use the resource ID and your API key to submit jobs to the Strangeworks platform. To see all the jobs submitted to that platform, check out the Related Jobs overview, shown on the left side of Figure 15-11. Scrolling down shows relevant billing transactions.

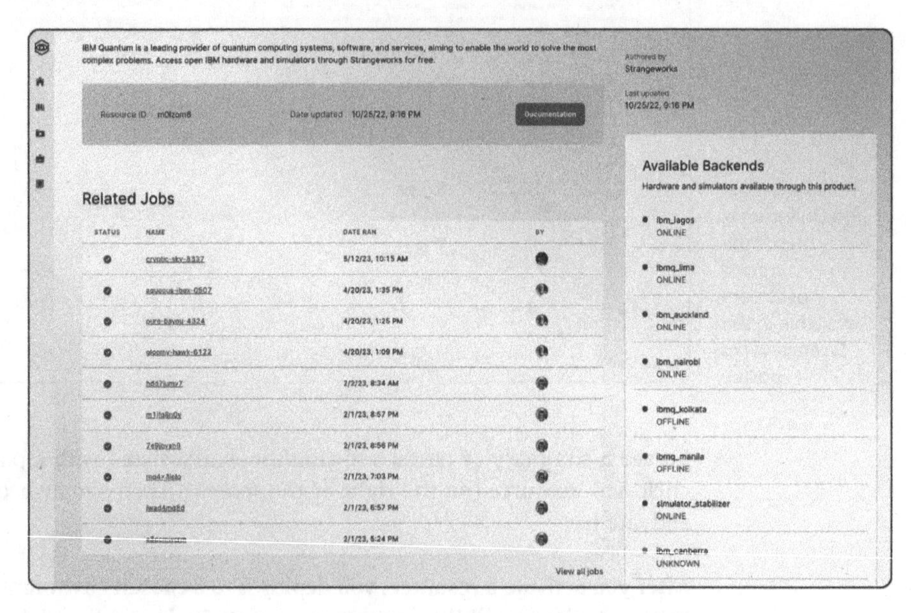

FIGURE 15-11: You can see jobs waiting to run on your chosen resource.

Figure 15-12 shows the screen that includes details for a specific job, including the circuits that were run, the histograms, and the results.

Understanding managed applications

Moving on from compute providers, further down on the dashboard page (refer to Figure 15-6) is a section for managed applications, as shown in Figure 15-13. *Managed applications* are slightly abstracted versions of products available to run on several of the compute providers available through Strangeworks. A good example is the Strangeworks optimization service, where with a single product you can submit QUBO formulations to multiple providers on the Strangeworks platform.

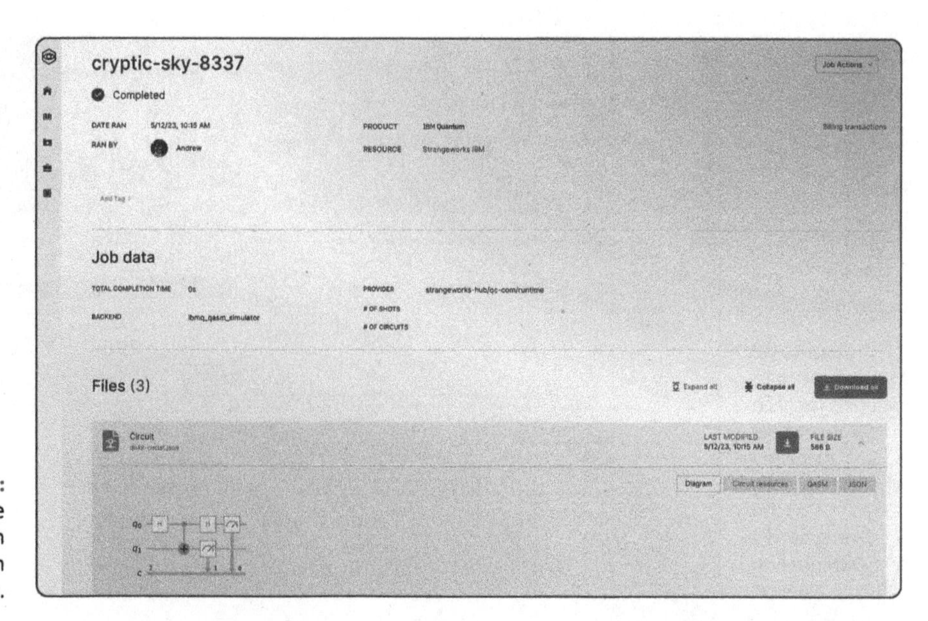

FIGURE 15-12: Click to see the details of each job running in your workspace.

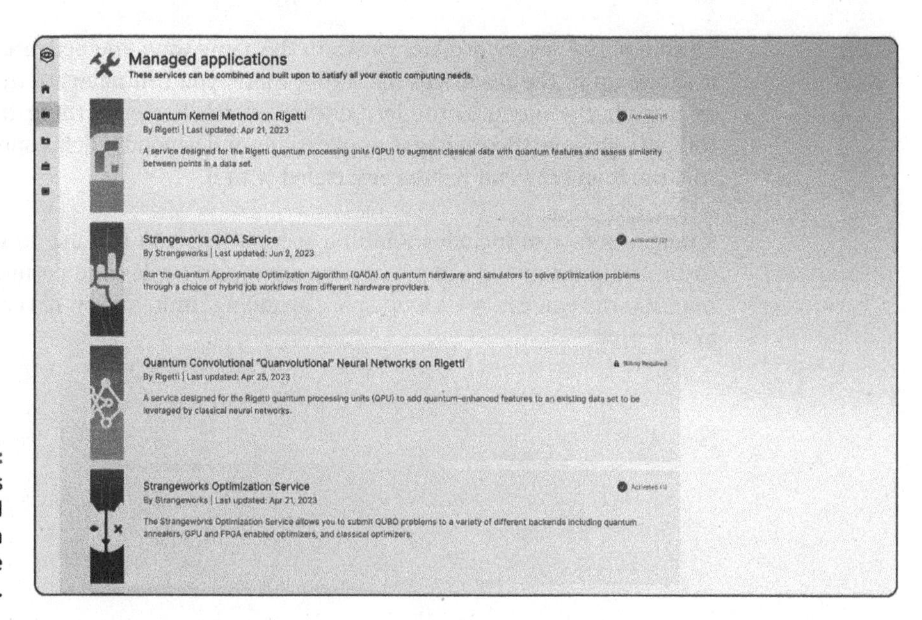

FIGURE 15-13: Strangeworks managed applications run on multiple providers.

Click to see details for a specific service. For instance, the Strangeworks Optimization Service, shown in Figure 15-14, is a popular service that runs across multiple providers. This gives you the ability to compare performance, cost, and more on different providers.

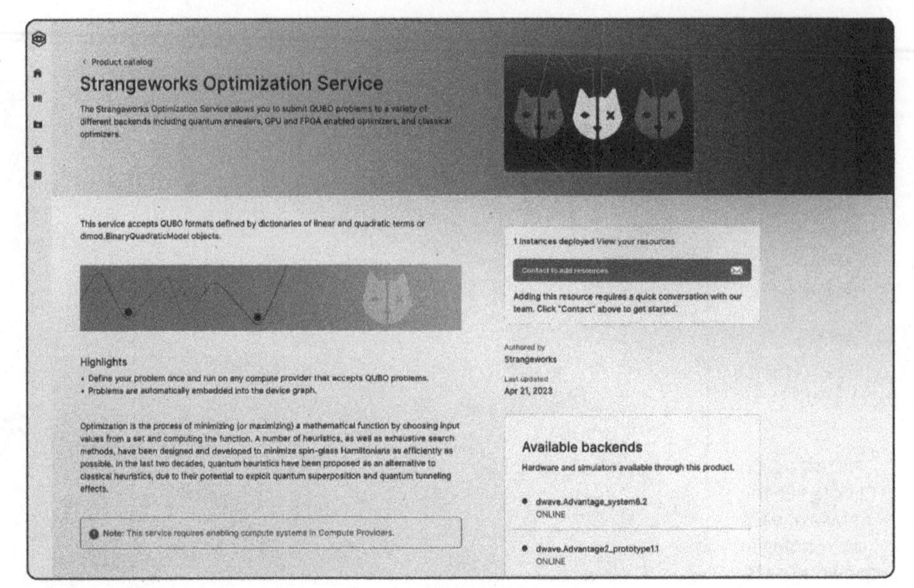

FIGURE 15-14: You can click into details for each Strangeworks managed application, such as the Strangeworks Optimization Service.

To summarize, every product works in the same way: You activate a product and it shows up in the resources list. From there, you can navigate to your resource catalog via the menu in the left sidebar, which has everything from individual compute providers to managed applications and beyond. Click a specific job to see who ran it and the end results associated with it.

Strangeworks also includes a billing section that you can use to keep on top of costs. And you can add users, granting appropriate roles and permissions to each. Importantly, you can set a workspace spending limit, so you never get a shock at billing time.

» Digging into the documentation

» Having a quantum of fun

» Taking a random walk around the web

Chapter 16 Educational Resources

So you want to program a quantum computer. Where does one start on such an epic journey?

A quick internet search will yield a wide range of online classes provided by universities, companies, and startups. But, you must be asking, which resources will be the most valuable to you on this journey?

Answering that question is precisely why we wrote this chapter. There are three options for learning how to program quantum computers: structured courses, unstructured courses, and courses that are just plain fun. In this chapter, we explore all of these. And since you've got a lot to take in on your quantum journey, let's get right to it.

Connecting with Online Classes

If you're interested in learning to program a quantum computer, an online class may be just what you need. This approach requires organizing and planning your study sessions, almost like returning to school.

Options such as tutorials and online classes can help you stay focused and motivated, improve your memory retention, and make the most of your time. You can achieve more efficient and effective study sessions by breaking down the learning materials into manageable chunks and setting specific goals. With this approach, you can also identify areas for improvement and track your progress over time.

THE BUSINESS OF QUANTUM COMPUTING

Many people are interested in applications for quantum computing from a developer's or competitive point of view. Others are interested in the industry view: where to invest and what the effect will be.

If you're interested in the industry view: Stop, do not pass go, and head to the Boston Computing Group (BCG) website, at `www.bcg.com/capabilities/digital-technology-data/emerging-technologies/quantum-computing`. They have a lot to say, and they do their homework, tying the emerging quantum computing sector to the broader economy.

Some classes may be more challenging than others, but by describing them here, we hope to help you gain the quantum knowledge you need to enter this exciting field.

MIT Quantum Information Sciences

`https://ocw.mit.edu/courses/8-370x-quantum-information-science-i-spring-2018/`

The online class for MIT Quantum Information Sciences, which features the image shown in Figure 16-1, offers an exceptional opportunity to delve into the fascinating field of quantum information and computation from the renowned Massachusetts Institute of Technology (MIT).

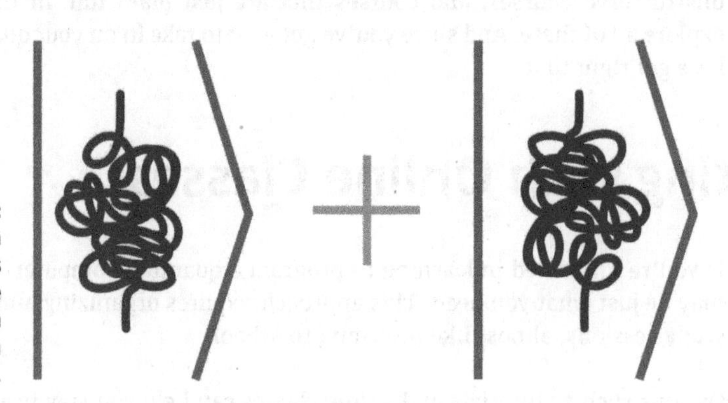

FIGURE 16-1: Quantum superposition, as rendered by the MIT Quantum Information Science site.

OPERATIONAL EXPERTISE, DEPTH, AND CREATIVE FREEDOM

The world has a *lot* of software developers — 4.4 million in the US alone, according to a recent estimate, which is a bit more than 1 percent of the population of the entire country. As in any other field of endeavor, an explosion of numbers leads to specialization. What skills and insights will you bring to quantum computer programming?

One contribution you might make is operational expertise as a programmer across different disciplines. Let's say you're a really good programmer in your language of choice, and you want to learn how to make those programs work in the quantum world. You can focus on a "shut up and calculate" approach: Your quantum computing code compiles, executes, and yields useful results on current quantum computer hardware and software.

Or you can focus on getting close to the hardware. In this approach, you'll dive deep into the interaction between code and the compiler, the underlying assembly language — in quantum computing, the nearest thing to this at present is OpenQASM (pronounced "open chasm," which is apt) — and the qubits themselves. You might become a true expert in programming in the quantum realm.

And our final suggestion is that you consider learning quantum computer programming so you can go up a level and contribute to planning for future quantum computer hardware and software. You'll be with Shakespeare in declaring that "there are more things in heaven and earth than are dreamt of in your philosophy" — and you may well have to fight your fair share of battles with the Horatios of this demesne.

To paraphrase a character played by the late, great Robin Williams in the movie *The Fisher King* (1991): "The challenge is not to figure out what to do; it's to figure out who you are, and be that." Figure out what you're good at, apply those skills to quantum computing, and let your vorpal blade go snicker-snack, as Lewis Carroll put it in the classic poem "Jabberwocky." (The words the poet uses in the poem are all made up, but somehow the reader believes they are able to figure out just what he means.)

The class, split up into three portions, covers a wide range of topics, including the principles of quantum mechanics, quantum algorithms, quantum error correction, quantum cryptography, and quantum simulation. Led by expert instructors with extensive research experience, the course provides comprehensive lectures, interactive demonstrations, and hands-on exercises to deepen understanding and proficiency in quantum information sciences.

Participants have access to cutting-edge research materials and resources, enabling them to stay at the forefront of this rapidly advancing field. The online format allows for flexible learning, accommodating students from various backgrounds and time zones.

Whether you are a researcher, student, or professional seeking to expand your knowledge and skills in quantum information sciences, this MIT online class equips those interested in the field with the tools and insights needed to explore and contribute to the future of quantum technologies.

Quantum Cryptography

www.edx.org/course/quantum-cryptography

The affordable Caltech online course "Quantum Cryptography" focuses on how quantum communication provides security guaranteed by the laws of nature — including how to use quantum effects, such as quantum entanglement and uncertainty, to implement cryptographic tasks with levels of security that are impossible to achieve classically. And learning how to use quantum effects for security will give you a unique perspective on how to use them for programming as well. The course runs for 10 weeks, requiring about 6-8 hours of hard work from students each week.

The Quantum Internet and Quantum Computers: How Will They Change the World?

www.edx.org/course/the-quantum-internet-and-quantum-computers-how-w-2

Delft University is a leading university in the Netherlands. Their online course "The Quantum Internet and Quantum Computers: How Will They Change the World?" is taught in English. It's a valuable resource for beginners seeking to understand the transformative potential of quantum technologies.

This six-week course provides a comprehensive introduction to the fundamental principles and applications of quantum computing and the emerging quantum internet. Through a series of engaging lectures and interactive exercises, participants gain a solid foundation in quantum mechanics, quantum information theory, and the principles of quantum computing.

The course explores topics such as quantum algorithms, quantum error correction, and quantum cryptography. Additionally, it delves into the concept of a quantum internet and its implications for secure communication and distributed quantum computing.

The course emphasizes real-world examples and case studies, offering insights into the practical applications and societal effects of quantum technologies. It's an invaluable resource for those who are new to the field and seeking to understand how these technologies will shape the future.

Understanding Quantum Computers

www.futurelearn.com/courses/intro-to-quantum-computing

The online course "Understanding Quantum Computers" offers a conceptual introduction to the fundamental principles of quantum mechanics and the motivation for building quantum computers. While it sticks to fundamentals, it also offers a brief overview of quantum computing hardware and the budding quantum information technology industry, according to the course's description. The class makes an effort to shy away from lots of mathematical equations in favor of graphical representations of key concepts, which might be helpful for beginners.

Quantum Quest

www.quantum-quest.org

Want to get into quantum computing development before heading to college? QuSoft has it covered. The Quantum Quest web class is a specialized educational program designed for high-school students interested in exploring the fascinating world of quantum physics and quantum computing.

This interactive web class, which primarily communicates via Discord, provides a structured curriculum that covers the foundational principles of quantum mechanics, quantum information, and quantum technologies. The course reaches out to younger people using (somewhat) simplified text and appealing graphics like the one shown in Figure 16-2.

Through engaging lectures, interactive demonstrations, and hands-on activities, students gain a solid understanding of quantum phenomena, including superposition and entanglement. The class also introduces students to quantum computing and its potential applications.

The Quantum Quest web class is tailored to make complex concepts accessible and engaging for high-school students, fostering their curiosity and enthusiasm for quantum science. By participating in this program, students have the opportunity to delve into cutting-edge research topics and gain valuable insights into the future of quantum technologies.

FIGURE 16-2: Quantum Quest provides a learning opportunity designed for high-schoolers.

Quantum Machine Learning

`www.edx.org/course/quantum-machine-learning`

"Quantum Machine Learning," produced by the University of Toronto, offers a hands-on introduction to quantum computing and quantum-enhanced machine learning, including a code repository. The class also introduces several quantum machine learning algorithms and implements them in Python, according to the course summary. This self-paced nine-week online course is expected to require 6 to 8 hours per week.

Quantum Computing: Less Formulas — More Understanding

`https://online.spbu.ru/quantum-computing-less-formulas-more-understanding`

St. Petersburg University's "Quantum Computing: Less Formulas — More Understanding" is yet another solid introductory course on quantum computing, and one that spends most of its effort focusing on concepts. The available syllabus for this free online class breaks down exactly what will be covered over the course of five weeks, and it's available in multiple languages.

EVERY DAY HE WROTE THE BOOK

Dr. Thomas Wong teaches quantum computing at Creighton University, where he is assistant professor of physics. His book, *Introduction to Classical and Quantum Computing*, was written as a text for his students. It's available for sale on Amazon (`www.amazon.com/Introduction-Classical-Quantum-Computing-Thomas/dp/B09QP2ML3P`) as well as for free, online, in PDF form (`www.thomaswong.net/introduction-to-classical-and-quantum-computing-1e3p.pdf`)! The book introduces both forms of computing, classical and quantum, and shows you how to program and run code in the IBM Quantum Experience.

Black Opal

`https://q-ctrl.com/black-opal`

"Go from zero background to programming real quantum computers" is the pitch for the Black Opal course from Q-Ctrl. While the promise seems a bit broad, Black Opal includes a lot of resources and aims to prepare people for real jobs, including a certification option. Be prepared to learn about "super-duper-position" and more. (As the authors of this book, we're afraid that Black Opal is trying to elbow us aside in the race to be humerus.)

Trying Tutorials and Documentation

Devs love to drive business types crazy by skipping past the website, brochures, and white papers that the business thoughtfully provides and going straight to the documentation. But going straight to the docs works in classical computing, and it works in quantum computing too. Here are some quantum computing documentation sites and tutorials that are likely to be worth your time.

Nielsen and Chuang

`https://workedproblems.wordpress.com/category/nielsenchuang`

Serving as a kind of appetizer, Worked Problems in Physics is a WordPress site that shares exercises focused on quantum algorithm problems. It's independent of any particular programming language, toolkit, or make of quantum computer. Think of it as a sort of "writing prompt" for quantum math nerds.

Documentation for Forest and pyQuil

https://pyquil-docs.rigetti.com/en/1.9/

Rigetti Computing provides an expansive tutorial for learning and using pyQuil via its Forest SDK. pyQuil is an open-source Python library, hosted on Github, that produces programs in the Quantum Instruction Language (Quil). Quil assumes that in the near term, quantum computers will operate as coprocessors and will work in conjunction with classical CPUs, so Quil is designed to execute on a Quantum Abstract Machine with a shared classical/quantum architecture.

Quil programs can run on a Quantum Virtual Machine (QVM), a cloud-based classical simulation of a quantum processor that can simulate various qubit operations, according to the tutorial's introduction. And if this is all new to you, the good thing about this documentation is that it provides some entry level information as well.

Documentation for Ocean

https://docs.ocean.dwavesys.com/en/stable/getting_started.html

Ocean software is a suite of tools from D-Wave Systems for solving hard problems with quantum annealers — the type of quantum computer that D-Wave makes (see Chapter 9). And D-Wave does not want to leave you feeling like "an astronaut in the ocean" when you use it. (The quoted text is the title of a rap song that seeks to provide consolation for those suffering from depression and other mood disorders.)

Ocean runs on D-Wave's cloud-based Leap integrated development environment (IDE) and provides reusable/disposable workspaces for running code from a personal GitHub repository or a collection of code examples that can then be modified, according to the company. The documentation comes with plenty of helpful examples and methods for getting started, regardless of your prior experience developing for quantum computers.

Documentation for Xanadu's Strawberry Fields

https://strawberryfields.ai/photonics/demonstrations.html

The photonic continuous-variable approach to quantum computation leverages the properties of light, specifically the continuous variables associated with the quadrature amplitudes of the electromagnetic field, to perform quantum information-processing tasks. Unlike discrete-variable approaches that use discrete quantum bits (qubits), this paradigm employs continuous-variable quantum systems, such as continuous-variable qumodes, which are manipulated through the operations of squeezing, displacement, and interferometry. These qumodes can represent information in an infinite-dimensional Hilbert space, offering the advantage of high-dimensional quantum states and continuous degrees of freedom.

Xanadu's Software Development Kit (SDK) provides a comprehensive framework for implementing quantum algorithms using continuous-variable quantum systems. It offers a user-friendly interface and a suite of tools that enable researchers and developers to simulate and experiment with various quantum computing tasks. The SDK includes libraries for simulating photonic quantum circuits, optimizing quantum algorithms, and accessing Xanadu's cloud-based quantum computing resources, empowering users to explore and exploit the potential of continuous-variable quantum computation.

As you might hope from the friendly name, the site includes clever graphics and animations. You'll be feeling the good vibrational excitations (with apologies to the Beach Boys and their famous song, *Good Vibrations*, which the Beatles themselves appreciated) when you visit this site, as shown in Figure 16-3.

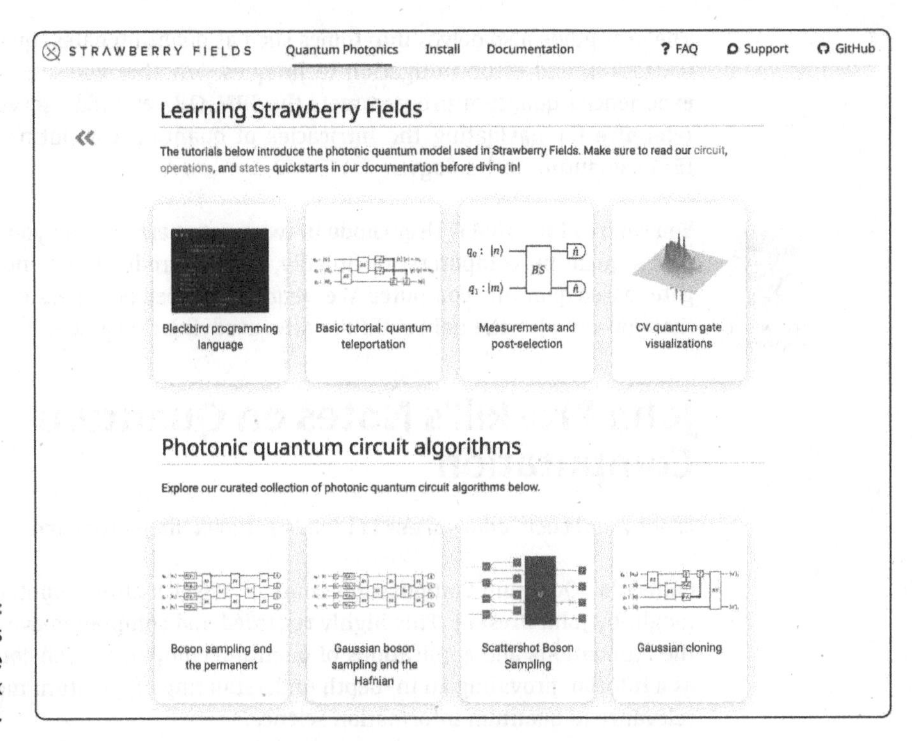

FIGURE 16-3: Strawberry Fields doesn't take forever to work through.

The name of the SDK is taken from a Beatles song, *Strawberry Fields Forever* (Lennon & McCartney, 1967). We can't promise that your work in Strawberry Fields will last forever, but we can promise that you'll learn a lot on the journey.

And Strawberry Fields isn't even the most popular SDK from Xanadu. Their biggest hit so far is PennyLane — which is indeed named after another Beatles song, *Penny Lane*. PennyLane helps you to optimize quantum circuits and to build hybrid models. The toolkit interfaces with machine learning libraries such as Keras, PyTorch, and Tensorflow.

IBM Q Full User Guide

`https://quantum-computing.ibm.com/lab/docs/iql/tutorials-overview`

The IBM Q User Guide is a comprehensive resource designed to provide users with the necessary information and guidance to effectively utilize IBM Q's quantum computing platform. The guide covers a wide range of topics, including an introduction to quantum computing, the principles of quantum circuits, and details on programming quantum algorithms using IBM Qiskit, a software development kit for quantum computing. It offers step-by-step instructions, code examples, and practical tips for writing and running quantum programs on IBM Q systems.

The user guide also delves into topics such as quantum gates, quantum state visualization, and error-mitigation techniques. Whether you are a beginner or an experienced quantum programmer, the IBM Q User Guide serves as a valuable reference for navigating the intricacies of quantum computing and leveraging IBM's quantum technology.

TECHNICAL STUFF

You can read the IBM Q User Guide in two ways: narrowly, so you can go program IBM's quantum computers; or broadly, as an introduction to programming any gate-based quantum computer. We suggest you keep both perspectives in mind so that you can play the field skillfully after studying this guide.

John Preskill's Notes on Quantum Computation

`theory.caltech.edu/~preskill/ph219/index.html#lecture`

Caltech's "Quantum Computation Course," Physics 219/Computer Science 219, is taught by John Preskill. This highly regarded and comprehensive course explores the foundations and applications of quantum computing. The course notes serve as a tutorial, providing an in-depth understanding of quantum mechanics and its relevance to quantum information science.

The notes delve into topics such as quantum algorithms, quantum error correction, quantum cryptography, and quantum simulation. Professor Preskill's expertise and engaging teaching style make the course accessible to both physics and computer science students, providing a solid foundation in the theoretical aspects of quantum computation.

The Preskill notes go into detail about how quantum communications between Bob and Alice might work, as shown in Figure 16-4.

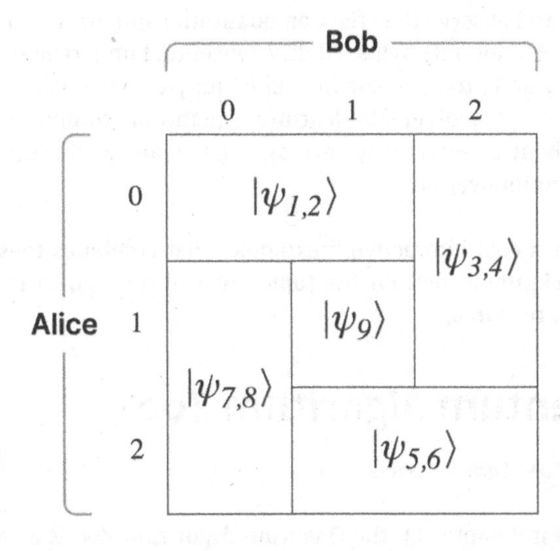

FIGURE 16-4: A mutually orthogonal interaction between Bob and Alice is described in John Preskill's Notes.

REMEMBER

As you may already know, Bob and Alice are the names usually assigned to two parties who want to communicate in a secure manner, including by using quantum communications.

The course also emphasizes practical aspects, with hands-on exercises and projects that allow students to implement and analyze quantum algorithms using various tools and platforms. The course is an excellent resource for anyone seeking a thorough understanding of quantum computation, bridging the gap between theory and practical implementation.

Qiskit Tutorials

`https://github.com/Qiskit/qiskit-tutorials`

Jupyter notebooks provide an interactive and educational platform for demonstrating how to use Qiskit, an open-source software development kit for quantum computing. Qiskit serves as a valuable resource for individuals learning quantum computing development. By following the step-by-step instructions and code examples in the notebooks, users can gain hands-on experience with Qiskit and explore various aspects of quantum programming.

The notebooks cover a wide range of topics, including quantum gates, quantum circuits, quantum algorithms, and quantum simulations. They also showcase real-world applications of quantum computing, such as quantum chemistry and optimization problems.

The interactive nature of Jupyter notebooks allows users to experiment, modify code snippets, and observe the effect on quantum computations in real time. This practical approach not only helps solidify conceptual understanding but also promotes creativity and problem-solving skills. Jupyter notebooks provide an invaluable resource for individuals learning quantum computing development, empowering them to effectively leverage Qiskit and embark on their journey toward quantum innovation.

IBM also offers a cool interactive "textbook" that combines these Jupyter notebooks with background text on the topic. Visit `https://qiskit.org/learn/` to access this cool resource.

The Quantum Algorithm Zoo

`https://quantumalgorithmzoo.org`

As we mention in Chapter 13, the Quantum Algorithm Zoo is a valuable resource for anyone learning quantum computing development. The Zoo serves as a repository of quantum algorithms, providing descriptions, explanations, and code implementations for each algorithm.

Its value lies in unbiased coverage of a wide range of quantum algorithms, including well-known ones like Shor's algorithm for factoring large numbers and Grover's algorithm for unstructured search, as well as more specialized algorithms for tasks such as optimization, machine learning, and cryptography. It offers a centralized and organized platform for individuals to explore and study various quantum algorithms.

As with the other resources listed in this chapter, the Zoo also provides a starting point for understanding the capabilities and potential of quantum computing, allowing developers to learn and experiment with different algorithms, gain insights into their underlying principles, and even contribute to the development of new algorithms. Ultimately, the Quantum Algorithm Zoo is an indispensable tool that promotes learning and innovation in the field of quantum computing development.

Quantum Computing Playground

www.quantumplayground.net/#/home

Google's Quantum Computing Playground is a web-based platform that offers an interactive and user-friendly environment for individuals seeking to learn quantum computing development, including a 3D quantum state visualization tool able to simulate up to 22 qubits. (Believe us, 22 qubits will be enough to expand your mind plenty.)

As its name implies, the Playground introduces you to a virtual playground where you can experiment with quantum circuits, algorithms, and simulations, all without the need for physical quantum hardware. This feature allows you to save those precious quantum computer credits for real work.

The platform offers a variety of prebuilt examples and exercises that cover fundamental concepts of quantum computing, including quantum gates, superposition, entanglement, and quantum teleportation. Users can manipulate and visualize quantum states, create custom quantum circuits, and observe the outcomes of quantum computations in real time.

The Quantum Computing Playground offers an intuitive interface that makes it accessible to beginners, as shown in Figure 16-5. It has a particular focus on Shor's algorithm (see Chapter 14).

The Playground also provides advanced features for more experienced users. It allows users to gain practical experience in quantum programming, explore quantum algorithms, and develop an intuition for the behavior of quantum systems.

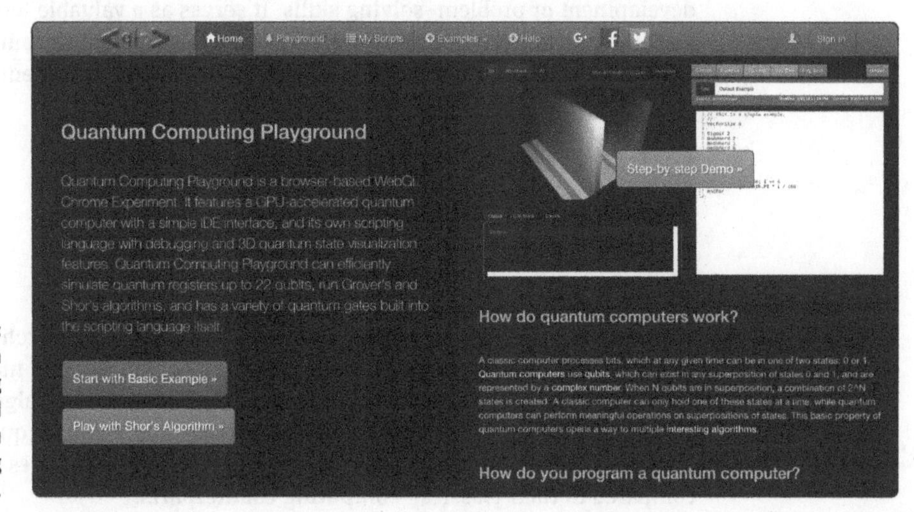

FIGURE 16-5: The Quantum Computing Playground makes Shor you learn something about the field.

By providing a hands-on learning environment, the Quantum Computing Playground serves as a valuable tool for individuals seeking to learn quantum computing development, enabling them to gain valuable insights and skills in a virtual quantum playground.

Quantum Katas

https://learn.microsoft.com/en-us/azure/quantum/tutorial-qdk-intro-to-katas

Quantum Katas is a valuable resource designed to support individuals learning quantum computing development by providing a collection of coding exercises and tutorials. Inspired by the concept of coding *katas* — simple exercises that you repeat many times to gain mastery — Quantum Katas offers a series of challenges that allow users to practice and refine their skills in writing quantum algorithms using Q# (Q-sharp), a programming language inspired by the classical-computing language C# but designed for quantum computing.

The katas cover a wide range of topics, including quantum gates, superposition, entanglement, and quantum algorithms such as Grover's search and Shor's factoring algorithm. Each kata provides a description of the problem, guiding questions, and a set of code templates that users must complete to solve the challenge. By working through the katas, individuals gain hands-on experience in quantum programming, learn to think "quantumly," and develop proficiency in implementing quantum algorithms.

The interactive nature of Quantum Katas, combined with detailed explanations and hints, facilitates a deep understanding of quantum concepts and fosters the development of problem-solving skills. It serves as a valuable tool for individuals learning quantum computing development, allowing them to enhance their practical skills and gain confidence in programming quantum systems.

Quantum Machine Learning for Data Scientists

https://arxiv.org/abs/1804.10068

"Quantum Machine Learning for Data Scientists," a research text by Dawid Kopczyk, explores the intersection of quantum computing and machine learning, focusing on the application of quantum machine learning algorithms in data science. It provides an overview of the current landscape of quantum machine learning techniques and discusses their potential advantages and challenges compared to their classical-computing counterparts.

It also examines various quantum machine learning algorithms, such as quantum support vector machines, quantum clustering, and quantum neural networks. This resource also discusses the effect of quantum noise, error correction, and the current limitations of hardware on the practical implementation of these algorithms.

Skipping through Unstructured Study

Unstructured study is one of the authors' favorite approaches; we did a lot of very unstructured study in pulling this book together. This mode of learning allows individuals to pursue their educational interests without the constraints of a rigid curriculum. It's especially beneficial if you have a strong desire to learn and to explore topics on your own terms. By pursuing unstructured study, you can learn at your own pace and delve deeply into the topics that most interest you.

Unstructured learning can contribute to a more fulfilling learning experience and a greater understanding of how to program a quantum computer. It can also help you learn "who's who" in the field as you browse different resources, so you can develop some connections to lean on as you move forward.

Additionally, unstructured study can help individuals develop skills such as self-discipline, critical thinking, and creativity, while applying these skills in the quantum realm. It may be particularly useful for individuals who have a passion for a particular subject but may not have the time or resources to pursue formal or structured education.

Blogs

To piggyback on the preceding paragraphs, one of the best ways to help connect the dots and remain up-to-speed in the field of quantum computing is by following dedicated blogs on the topic. Beginners may struggle to comprehend some of the more dense discussions and essay subjects, but blogs should be helpful to anyone just starting out to see what topics are trending among the active voices in the field.

Beyond that, quantum computing blogs are also a great way to advance your understanding over time, once you've taken some courses or started to build programs for a quantum computer. Keeping up with blogs is really one of the best ways to stay on top of things, which is why we've put together a comprehensive list here.

Algorithmic Assertions

https://algassert.com

The Algorithmic Assertions blog focuses on quantum computing and computing in general. It is authored on a time-available basis by programmer-turned-research-scientist Craig Gidney, a prominent member of Google's quantum computing team and creator of drag-and-drop quantum circuit simulator Quirk.

Bits of Quantum

https://blog.qutech.nl

Another great resource is Bits of Quantum, a blog offering occasional posts about quantum computing by members of research group QuTech Institute. While the volume of activity on this blog might be lower than you'd expect, it does offer a glimpse into ongoing research and daily life.

Decodoku

https://decodoku.medium.com

Decodoku is an occasionally updated blog that focuses on interesting posts about quantum computation. The site is authored by IBM Research staff member James Wootton, who also stays active on social media in-between articles.

Musty Thoughts

www.mustythoughts.com

Musty Thoughts is the personal blog of Polish quantum software engineer Michał Stęchły at PsiQuantum and, before that, Zapata Computing. While it doesn't get updated frequently, there are some decent essays and posts worth chewing on.

Quantum Frontiers

https://quantumfrontiers.com

One of the most accessible and regularly updated blogs on our list is Quantum Frontiers, which averages at least one to three posts per month. It's produced by the Quantum Institute for Quantum Information and Matter at Caltech, with a focus on sharing behind-the-scenes research insights.

Quantum Weekly

https://quantum-weekly.com

The pace of research breakthroughs and innovation in quantum computing moves quickly. It's easy to miss great new posts or articles even if you subscribe to everything else just listed. The Quantum Weekly blog offers a reliable recap of all things quantum — computing, cryptography, entanglement, and more.

Shtetl Optimized

https://scottaaronson.blog

Rounding out our list is Shtetl Optimized, a regularly updated personal blog authored by the very well-informed and highly opinionated Scott Aaronson. He is Schlumberger Centennial Chair of Computer Science at The University of Texas at Austin and director of its Quantum Information Center. Posts focus on quantum computing as well as AI, and other topics that intersect with the two.

Papers

Due to the rapid pace at which the quantum computing field moves, the content of a book can quickly become outdated. Thus, scientific articles are a good resource for keeping up with the state of the art.

Different versions of scientific articles are now far more available than they were in the past, often for free, though there's usually a charge for final versions in prestigious journals.

If you don't quite understand something in a scientific paper, usually a more accessible version is available in the scientific and technical press. Just search online for recent articles that include the name of the paper's author(s).

Opportunities and Challenges for Quantum Machine Learning

https://arxiv.org/abs/1708.09757

In the paper titled "Opportunities and Challenges for Quantum Machine Learning," researchers discuss what they would do if they had a quantum computer with 1,000 "perfect" qubits. Several examples are given about how quantum computers might excel in certain tasks in the field of machine learning. (As you probably know by now, *machine learning* is when a computer learns how to do something without being told in detail what to do; this approach seems well-suited to the unique capabilities of quantum computers.) It also introduces a new concept known as the quantum-assisted Helmholtz machine (QAHM) — so if you feel stressed about having so much to learn, you can Keep QAHM and Carry On. But seriously, this new technique incorporates both traditional and quantum computing to enhance learning capabilities, which is crucial to understanding the potential of quantum computing's capability to revolutionize machine learning in the near-future.

Quantum Machine Learning . . . and More

https://books.google.com/books/about/Quantum_Machine_Learning.html?id=92hzAwAAQBAJ&source=kp_book_description

"Quantum Machine Learning: What Quantum Computing Means to Data Mining" by Peter Wittek is a comprehensive exploration of the intersection between quantum computing and machine learning. The paper, now available in book form, delves into the potential of quantum computers to revolutionize data mining and presents a comprehensive overview of the field.

The paper covers a wide range of topics, including the principles of quantum mechanics, quantum algorithms for machine learning tasks, and the advantages and challenges of implementing quantum machine learning models. Wittek provides practical insights into quantum-inspired algorithms and explains how they can be applied to solve complex optimization and pattern recognition problems. It also discusses the implications of quantum machine learning for various industries, highlighting the potential for enhanced data analysis and decision-making capabilities.

The paper not only explains the fundamental concepts of quantum mechanics and quantum computation but also provides practical examples and case studies to demonstrate the application of quantum machine learning techniques. With its accessible writing style and comprehensive coverage, the paper (which is closer in size and impact to a book) serves as a valuable resource for researchers, practitioners, and students seeking to understand the implications of quantum computing on the field of data mining and machine learning.

Making Quantum Computing Open: Lessons from Open-Source Projects

www.researchgate.net/publication/330870969_Making_Quantum_Computing_Open_Lessons_from_Open-Source_Projects

The research paper "Making Quantum Computing Open: Lessons from Open-Source Projects" explores the importance and benefits of open-source initiatives in the field of quantum computing. It does this by examining lessons learned from successful open-source projects in the broader context of software development and applying them to the emerging field of quantum computing.

The paper highlights the advantages of openness, collaboration, and transparency in driving innovation, fostering community engagement, and accelerating the development of quantum technologies. Additionally, it discusses potential challenges and provides recommendations for establishing effective open-source practices in the quantum computing domain.

Getting into Interaction and Fun

When it comes to learning quantum computing, interaction and fun can be incredibly effective tools. Making things enjoyable and engaging can help learners to retain information more easily, as well as encourage people to ask questions and explore concepts more deeply. Whether through hands-on activities, games, or group discussions, finding ways to make quantum computing fun can help learners stay motivated and focused on their learning goals. So if you're looking to learn about quantum computing, don't be afraid to embrace the fun and interactive side of learning — it could be the key to unlocking your full potential!

Communities

One of the best, most underrated forms of engagement for complex topics such as quantum computing is online community forums. What's more, a plethora of quantum-related platforms are out there, providing a natural next step to staying engaged with whatever aspect or code base you settle into when getting into quantum computing.

Following is a quick roundup of communities that should be on the radar for anyone new to the field:

- **IBM Q Community:** IBM Quantum Experience Community Forum.
- **IBM Q QISKit Community:** Slack Channel for Qiskit and quantum computing discussions.
- **Q Community:** An open-source community around quantum programming in Q, including blogs, code repositories, and online meetups.
- **Women in Quantum Computing and Applications:** An inclusive meetup group based in Seattle, WA, US (currently meeting remotely) that hosts talks and tutorial series.
- **Mike & Ike Subreddit:** Discussion about the book *Quantum Computation and Quantum Information.*
- **Quantum Computing StackExchange:** Question-and-answer site for quantum computing. It's perfect for seeking answers to specific questions not

found in the available research papers and other quantum computing resources. Plus, it puts you in with other quantum computing geeks!

- **Quantum Information and Quantum Computer Scientists of the World Unite:** Facebook group for quantum research discussion.
- **Rigetti Community:** Slack Channel for Rigetti and quantum computing discussions.

Also please consider Meetups. Meetups used to be focused on IRL (in real life) meetings; however, since the pandemic, many are virtual, allowing you to learn from experts and connect with others all around the world.

TIP

You have to mess with Meetup's distance filter to see groups outside your area, but it's worth the effort.

One quantum computing Meetup group that's worth trying from any time zone is the Washington, DC Quantum Computing Meetup, which is near both America's seat of government and the headquarters of the nation's security agencies. Go to www.meetup.com/washington-quantum-computing-meetup. Keep an eye out for questions about cryptography from audience members who don't carefully identify themselves.

Interactive Learning Tools

We highly recommend these tools, because interactive learning tools can be fun! That makes them a nice break from the more po-faced resources above.

Quirk

https://algassert.com/2016/05/22/quirk.html

Quirk is a quantum circuit simulator that allows users to design, visualize, and analyze quantum circuits in a user-friendly and intuitive manner. This web-based simulator provides a drag-and-drop interface, making it accessible even to beginners in the field of quantum computing.

Quirk supports various quantum gates and operations, enabling users to construct complex quantum circuits by simply dragging and connecting elements on the canvas. It provides real-time visualization of the quantum state and allows users to observe the effects of different gate sequences on qubits. The site includes a quantum circuit editor, as shown in Figure 16-6.

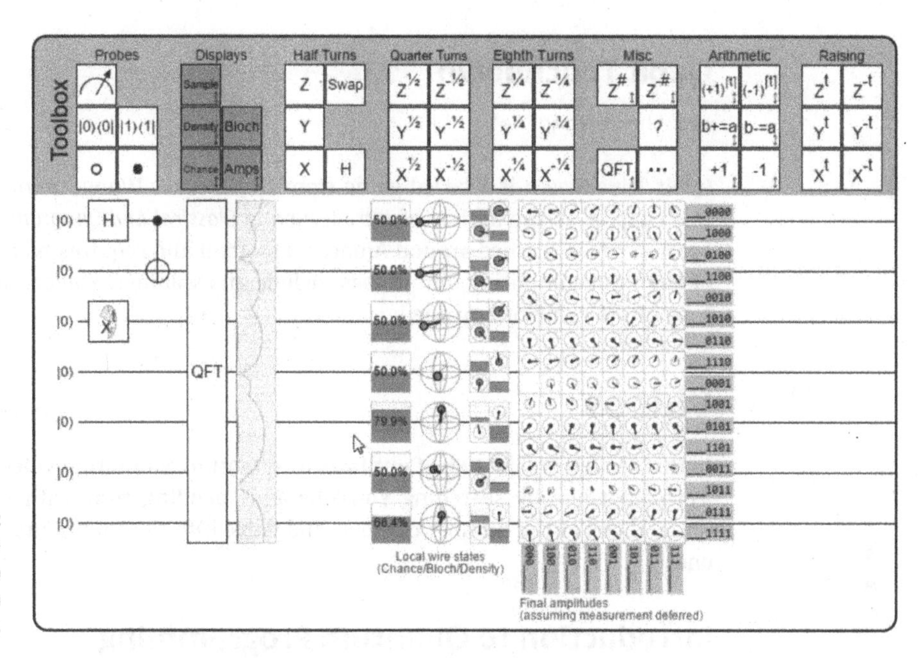

FIGURE 16-6: Quirk may have its quirks, but so does quantum computing.

Quirk also offers features such as measurement, state visualization, and noise simulation, providing a comprehensive toolkit for exploring and understanding quantum circuits. With its simplicity and interactive nature, Quirk serves as an invaluable learning and experimentation tool for individuals seeking to grasp the concepts and behavior of quantum circuits.

Quantum Odyssey

`www.quarksinteractive.com/download-quantum-odyssey/`

Quantum Odyssey is an educational game for Microsoft Windows that takes players on an exciting journey through the world of quantum physics and quantum computing. The game combines interactive gameplay with educational content to make learning about quantum concepts engaging and fun. Players embark on quests and challenges that require them to apply quantum principles and solve puzzles using quantum mechanics. Through gameplay, players gain a deeper understanding of quantum phenomena such as superposition, entanglement, and quantum algorithms.

Quantum Odyssey offers an immersive and interactive learning experience, making it a valuable tool for students, educators, and anyone interested in exploring the wonders of quantum physics in an interactive and gamified environment.

Qubit Touchdown

www.thomaswong.net

Qubit Touchdown is a board game from Dr. Thomas Wong, mentioned earlier in this chapter as the author of *Introduction to Classical and Quantum Computing*. His board game is themed around American football and requires no prior knowledge of quantum computing, but it does include an explainer. You can make your own copy of the game or buy one.

Videos

In this section we list two of our favorite quantum computing videos. You can also learn a lot just by searching YouTube and spending time with videos you find there. (Look at the number of views and delve into the comments to find the best ones.)

Introduction to Quantum Programming

https://skillsmatter.com/skillscasts/11929-programming-the-world-s-first-quantum-computers-using-forest

"Introduction to Quantum Programming" (duration 1 hour and 22 minutes) covers why you might want to program a quantum computer and how you would do so today, with a focus on the Forest Python SDK from Rigetti Computing. The only background assumed is linear algebra and complex numbers at the level of undergraduate computer science.

Quantum Computing for Computer Scientists

www.youtube.com/watch?v=F_Riqjdh2oM

"Quantum Computing for Computer Scientists" (duration 1 hour and 28 minutes) is a Microsoft Research Talk on introductory quantum computing for computer scientists. It stays away from pop culture sci-fi metaphors in favor of focusing on a single important question: "From a computer science perspective, how can a quantum computer outperform a classical computer?"

The video educates viewers on representing computation with basic linear algebra (matrices and vectors); the computational workings of qubits, superposition, and quantum logic gates; solving the Deutsch oracle problem (the simplest problem where a quantum computer outperforms classical methods); and tackling quantum entanglement and teleportation.

4
The Part of Tens

IN THIS PART . . .

Review quantum computer success in the early days.

Pursue answers to tech-side questions from the floor of presentations by whurley.

Browse business-related questions from the floor of whurley presentations.

Investigate leading university research programs in quantum computing.

» Debunking the need for a physicist to do everything

» Pointing out the demand for more talent

Chapter 17
Ten Myths Surrounding Quantum Computing

The topic of quantum computing has stirred up a lot of excitement and speculation in the tech world. And not just experts are fascinated by this field; the general public is intrigued by its potential as well. Quantum computing has been touted as having unparalleled computational power that could lead to groundbreaking advances.

However, all this hype comes with myths and misconceptions, creating a fog of FUD (fear, uncertainty, and doubt). The hype has also generated a boatload of misunderstandings surrounding quantum computing.

In this chapter, we aim to call out the top ten myths, put them to rest, and bring to light the realities of this revolutionary field. By dispelling these falsehoods, we hope to help people gain a more accurate understanding of quantum computing's potential, limitations, and likely effect on our future.

TECHNICAL STUFF

In this chapter and throughout the book, when we refer to a *qubit*, we mean a qubit in a universal, gate-based quantum computer, as described in Chapters 10 and 11. Quantum annealers, described in Chapter 9, also use qubits, but they don't have the full power of the qubits used in gate-based quantum computers, and they aren't used in the same kind of processes.

Myth 1: Quantum Computing Won't Be Commercially Available for 10–15 Years

Quantum computers have long been shrouded in a veil of mystery, with speculation about their commercial viability dominating conversations. As interest and investment in this field continue to grow, the number of hyperbolic news headlines about quantum computing's potential have contributed to the notion that it's still closer to science fiction than reality. However, the belief that it will take another 10 to 15 years for quantum computing to become available is, we believe, a myth.

Quantum computers are commercially available. You can access them via the cloud today, and they are already being used by researchers, government agencies, and corporations. Major technology companies such as IBM, Google, and Microsoft have made significant strides in developing and offering their quantum computers to the public. Additionally, startups like D-Wave Systems, IonQ, and Rigetti Computing have actively made quantum computing commercially available.

In addition, providers like AWS Braket, Microsoft Azure, and Strangeworks offer users access to quantum computers, code samples, and educational resources via the cloud. Most importantly, the lion's share of these quantum startups — and yes, even the big guys are startups in the quantum realm — are making money. Albeit not a lot of money, but enough to definitively declare that quantum computing has gone commercial. (More proof: Long-time insiders in the field already miss the days when it hadn't yet done so.)

What is true, however, is that quantum computing is not yet commercially *successful*. We do not have real-world applications of quantum computing that are changing the way people do things and making lots of money for their backers.

SO MANY MYTHS, SO LITTLE TIME

We chose to address these ten myths because we heard them a lot while working on this book. It's important to remember that quantum computing is an exciting field with numerous possibilities to address world-changing challenges that have thus far eluded humanity. But quantum computing is still emerging, so it's vital for all involved, and those who want to get involved, to recognize the many common quantum computing misconceptions. By dispelling these myths and gaining a clearer understanding of the realities and potentials of quantum computing, we can foster more informed discussions and harness the transformative power of this emerging technology.

There are also still many unanswered questions about the technology. Quantum-inspired computing, running on classical hardware, is starting to rack up some successes (as described in Chapter 8), but quantum annealing, which is used for the same kind of optimization problems, continues to try to prove itself (see Chapter 9). And gate-based quantum computers are just gaining traction (see Chapters 10 and 11), but they still need technical advances, including many more qubits and successful error-correction technology, to have a large chance of success.

The activity of prestigious organizations, coupled with ongoing research, suggests that quantum computing will achieve real-world breakthroughs much sooner than anticipated. We believe that the quantum computers now becoming available are laying the groundwork for commercial success.

Myth 2: A Qubit Can Be a 0 and a 1 at the Same Time

One of the fascinating aspects of quantum computing we discuss in the book is the concept of superposition, which allows qubits to exist in multiple states simultaneously. However, one misconception is that a qubit can represent 0 and 1 simultaneously. This common myth may make it more difficult to understand the vast possibilities quantum computing offers.

In reality, a qubit can be in a superposition of states, meaning it represents the value for a linear combination of real and imaginary numbers that changes during a computer program run (called a circuit). This flexibility, combined with the capability of qubits to be entangled with each other, enables quantum computers to perform parallel computations, resulting in exponential speedups for certain algorithms. Nevertheless, when measured at the end of the circuit, the qubit collapses into a single state, which is measured as either 0 or 1.

Language is messing us up here. In the quantum world, which qubits are part of during processing, we can't measure a quantum value without ending the process we're measuring. The qubit has no defined value until it's measured — and measuring ends the program run. So you begin by putting the qubit in an unknown state; you manipulate it, quite precisely, using programming steps, but still without knowing its value (because it doesn't have one); and finally, you stop the program run and measure the value.

A more accurate version of the statement in myth 2 might be to say the following:

- » A qubit is first initialized to a value that could, if it were measured, result in a 0 or a 1 being returned.
- » The qubit is then programmed through a series of steps, called gates. At each of those steps, if the qubit were measured, it might return either a 0 or a 1.
- » When all the gates are completed, the qubit is measured, and finally returns either a 0 or a 1.

It can be difficult to conceptualize a value of superposition because it seems incompatible with how we conceptualize binary values for classical computing. But clarifying this misunderstanding, to the extent that we can, is crucial to better comprehending the potential of quantum computing.

Myth 3: Quantum Computers Will Replace Classical Computers

The notion that quantum computers will entirely replace classical computers is a commonly repeated myth and a misconception at best . . . unless we're looking one hundred years into the future, at which point, who knows? But for now, quantum computers excel at solving specific types of problems; they're not designed to replace classical computers outright. They also don't need to, because each is better for different kinds of tasks. Put another way, using a screwdriver to drive in nails, rather than a hammer, doesn't make much sense — even if it is possible, albeit far more challenging and potentially dangerous.

Classical computers remain vital for the tasks we use them for today, including everyday computing needs, general-purpose programming, moving data around, and non-quantum algorithmic problems found in high-performance computing (HPC) and supercomputing.

Quantum computers, on the other hand, offer unique advantages for solving complex optimization problems, factoring large numbers, simulating quantum systems, and cryptography. You would not, for example, use a quantum computer to conduct basic arithmetic that you do using Excel.

If you need a truly random number in Excel, a classical computer can't produce one. The RAND() function in Excel actually generates what's called a pseudorandom number, not a truly random one; the device is inherently incapable of real randomness. For a truly random number, fire up a qubit. (Or you can watch a butterfly and count whether it flaps its wings an odd or even number of times over, say, a minute. But you can't always count on having a butterfly nearby when you need one.)

The classical computing paradigm and the quantum computing paradigm are expected to coexist, each serving different computational needs, forever. Moreover, having access to quantum computers (and the solutions to problems the they're uniquely suited to solve) through classical computers, which will deliver data to the quantum computers and receive and process the results, will be crucial.

Myth 4: Only a Physicist Can Program Quantum Computers

It's commonly believed that only physicists possess the knowledge and skills to program quantum computers. The number of sci-fi films and television series with brilliant mad scientists applying quantum technology tends to reinforce this belief. However, this myth is becoming less true every day, especially given the steady progress made in democratizing access to quantum computing — progress that is often overlooked.

With the development of user-friendly programming languages, software frameworks, comprehensive documentation, and improving abstraction layers that somewhat hide the machines' complexity, programming quantum computers is becoming accessible to a broader audience. For instance, companies such as IBM, Microsoft, and Strangeworks offer cloud-based platforms equipped with intuitive interfaces and tools, empowering users from various backgrounds to engage with quantum programming in a way that will foster greater collaboration and innovation. As quantum computing evolves, more resources and educational initiatives are emerging, ensuring that programming quantum computers will not be limited to physicists alone.

Myth 5: Quantum Computers Will Soon Solve All Classical Computer Problems

Quantum computers offer the potential for extraordinary computational power, but the belief that they will soon solve all classical computer problems at once is a misconception. We're not sure where this myth started, but don't let one of the co-authors (whurley) find out, because he is unhappy.

While quantum algorithms can deliver exponential speed boosts for certain problems, they don't provide a universal solution for all computational challenges. Quantum algorithms — such as Shor's algorithm for factoring large numbers, Grover's algorithm for unstructured search, and quantum simulation algorithms — offer significant advantages over classical counterparts. However, classical computers are more efficient for solving many other problems, including most everyday computational tasks. Recognizing the limitations and scope of quantum computing is essential in helping all involved to avoid unrealistic expectations.

What is true is that quantum computers will solve more complex problems over time. Today's quantum computers can't run Grover's algorithm on large enough searches to be very useful, but they might in a few years. And they can't run Shor's much more challenging algorithm on large enough prime numbers to be very effective, and getting to that point may take more than a decade. A better painkiller for your headache might arrive a few years from now; a new and improved head that never gets headaches, though, might take a while.

Myth 6: We Should All "Shut Up and Calculate"

There comes a time in any human endeavor when it's time to stop fretting, figuring, and fine-tuning, and instead pound poles into holes until the job is done. So it is with quantum computing, and the popular phrase, "shut up and calculate," captures that spirit nicely.

However, some people reflexively use this phrase whenever the deeper mysteries of quantum computing and quantum mechanics come up, and we believe that is a mistake. There's a need for a laserlike focus on execution at times, but there's also a strong need to pause the calculating at times and let your mind roam.

Every field needs dreamers and all of us need to include dreaming in our approach — and this has never been more true than in quantum computing.

Think big. Start by getting outside and going for a walk. Play with a dog, talk to a child — or vice versa — and free up your mind. Then look through Chapter 3, on quantum mechanics, or page through the applications listed in Chapter 13 and, if you're technical, the algorithms listed in Chapter 14. Imagine new ways to think about the problems you're trying to solve in the quantum computing field and in your work more broadly. We'd be surprised if you don't come up with some novel answers.

Myth 7: Soon There Will Be Only a Small Number of Quantum Hardware Companies

The belief that a small number of companies will dominate the quantum computing industry is, we believe, a myth, perpetuated in part by the idea that only organizations with the means to produce quantum computing hardware at a massive scale will be able to succeed, or even to continue to operate.

The quantum computing landscape is characterized by a diverse array of hardware providers, ranging from established tech giants to innovative startups. Companies such as D-Wave Systems, Google, IBM, IonQ, Microsoft, and Rigetti Computing are actively engaged in quantum hardware research and development. (Many of these companies, and others, also provide software and algorithm services.) These players don't operate in a silo or a vacuum, either. Instead, there's a legitimate need to forge partnerships to provide access to hardware that ultimately helps drive innovation more efficiently and effectively.

Quantum computing is an ever-changing, vibrant ecosystem that will feature hundreds, if not thousands, of companies. And as the field progresses, we can expect to see increased competition between a growing number of players, fostering innovation and driving the industry forward.

Myth 8: Quantum Companies Have All the Talent They Need to Grow the Industry

While some significant figures in the technology — – often appearing in pairs, such as Bill Gates and Steve Ballmer at Microsoft, "the two Steves" at Apple — the late Steve Jobs and Steve Wozniak — and Mark Zuckerberg and Sheryl Sandberg at Facebook — possess a combination of technical and business acumen, it's a myth to assume that all technical and business leaders needed for quantum companies are already in place and hard at work. If that were the case, the prevalence of quantum computing solutions would be far more prominent by now. This sector is still emerging, constantly identifying new roles and unique skill sets that are needed as the discipline pushes forward.

Quantum computing is a rapidly evolving field that requires interdisciplinary expertise, including physicists, computer scientists, mathematicians, engineers, and business professionals. The demand for skilled talent in quantum computing far exceeds the current supply, making it an ongoing challenge for companies to attract and retain top talent. Collaborations between academia and industry, educational initiatives, and investment in talent development are essential to nurturing the growth of the quantum computing industry.

Myth 9: Quantum Computing Will Destroy Data Encryption

The notion that quantum computing will spell the end of encryption and cybersecurity is a myth that can be dispelled quickly. It's true that quantum computing may, in many years, be able to defeat today's data encryption standards. But, if quantum computing can break today's best encryption methods, it can also be used to build new, more secure solutions.

Security is an ever-entwining dance of threat and remediation. No one needs the Enigma decryption machine from World War II anymore, nor does any nation use any encryption that's not the latest and greatest for its most important secrets. So quantum computing is just the latest challenge that will inspire us to improve the rigor of our security standards and also give us new means to do so.

While quantum computers have the potential to break currently used cryptographic algorithms, such as RSA and ECC, the development of quantum-resistant encryption methods, also known as post-quantum cryptography, is underway.

Researchers worldwide are actively exploring alternative, quantum-powered cryptographic schemes that can withstand attacks from quantum computers (and also, one hopes, aliens). Governments, research organizations, and experts are collaborating to develop and standardize post-quantum cryptographic solutions to ensure the long-term security of sensitive data.

The challenge here is real: If quantum computing reaches the point where it can beat today's cybersecurity, information protected with current methods will be exposed. Some ill-intentioned folk are saving copies of today's encrypted documents in the hope of using quantum computers to expose their contents in the not-so-distant future.

So it's essential to stay vigilant and proactive in adopting quantum-resistant encryption methods to maintain cybersecurity in the era of quantum computing. But to say that quantum computing will destroy cybersecurity indefinitely is shortsighted.

Myth 10: Quantum-Safe Cryptography Provides Complete Data Security

Given time, resources, and enough patience, no lock is impenetrable. The same is true for the capability to protect data from all threats. While solutions like quantum-safe cryptography offer resistance against attacks from quantum computers, it's a myth to believe that it provides complete data security. (Quantum-safe cryptography refers to cryptographic algorithms resistant to attacks from classical and quantum computers.)

While these algorithms are expected to mitigate threats posed by quantum computers, they're not immune to all potential attacks. Cryptographic systems are complex, and vulnerabilities can emerge over time. Therefore, a layered approach to cybersecurity, combining quantum-safe encryption, secure critical-management practices, and ongoing monitoring of emerging threats, is crucial to ensuring comprehensive data security in the quantum computing era.

need him

» Estimating the arrival date for practical quantum computing results

» Anticipating the results when quantum computing meets AI

Chapter 18
Ten Tech Questions Answered

One of the authors (whurley) gives a lot of quantum computing talks at conferences and other events around the world. The following questions about quantum-based technologies in general, and quantum computing in particular, are from recent talks. This chapter focuses on questions about the technology, starting with a couple of questions about quantum mechanics in general and followed by questions about quantum computing. See the next chapter for a related set of questions about the business of quantum computing.

Will Quantum Technology Find Its Way into a Consumer Product?

The basic principles of quantum mechanics were set out between 1900 and 1930, as described in Chapter 3. But practical uses of quantum technology followed quickly, including television (1927), the atomic bomb (1945), and lasers (1960).

Questions important to quantum computing have been considered for centuries, but the first proposal for a quantum computer was put forth by Richard Feynman in the early 1980s. However, quantum computers today are mostly large, expensive, and supercooled to close to absolute zero. As the questioner implies, these systems are even harder to build, install, and manage than early mainframe computers, and neither kind of system is a consumer product.

Today's quantum computers are mostly accessed via the cloud, with users paying for use on a timesharing basis. This method of access will probably be dominant for years to come. What would it take for quantum computing technology — qubits, in particular — to be included in a personal computer or smartphone?

The simplest use case we can imagine is a true random-number generator that uses a single qubit to encapsulate the quantum randomness needed to generate truly random numbers, independent of any outside source or interference. Further down the road, one or two qubits could be used in devices as part of a quantum-secured communications system.

We don't currently see a need for quantum computing (that is, multiple qubits) in a consumer product, but never say never. (And yes, someone will probably quote this sentence and laugh some years down the road, after quantum computers start being sold to students at college bookstores.)

Is the Quantum Realm Real? Will Ant-Man Save Our World?

The questioner is referring to quantum mechanics in general and to several recent movies on the topic, all released in 2022: *Ant-Man and the Wasp: Quantumania*, *Everything Everywhere All At Once* (referred to in the trade as EEAAO), and *Doctor Strange and the Multiverse of Madness*. Quantum technology also featured in the Amazon Prime Video series *The Man in the High Castle* and the 2023 movie *Oppenheimer*, and advanced quantum technology is also expected in the upcoming (as this is written) Netflix series *3 Body Problem*.

We love seeing movies such as these being produced because they're part of the process by which quantum mechanical thinking becomes popularized. We expect that it will be people born in the near future, in a world where quantum mechanical principles and quantum computing are commonplace, who will take the technology to places we can't imagine today.

And yes, some of those applications may require Ant-Man, or at least some kind of heroic effort, to save us. Quantum mechanics already enabled the atomic bomb, which is definitely a danger to humanity. (One can, far less seriously, say the same thing about television.)

In the case of quantum computing, we mention in Chapter 13 how we see quantum computing as vital to the future development of AI and machine learning, which is another technology that can fuel dreams as well as nightmares. (1970's *Colossus: The Forbin Project* and *The Terminator*, which launched in 1984 — appropriately — are among many examples.) So we need heroes and superheroes both to move technology forward and to protect us all from the technology being misused.

How Do You Explain Quantum Computing to a Dummy?

Quantum computing uses qubits, which have quantum mechanical qualities. That means they act in strange ways that are visible at the level of atoms, electrons, and photons but not in our everyday lives. Each qubit exhibits superposition, the capability to hold many values at once. Two or more qubits can be connected to each other via entanglement, in which seemingly independent quantum particles take on aspects of one another's state.

As quantum computing advances, qubits will be able to run algorithms that bring previously unheard-of power to tasks such as optimizing multistep processes and multi-actor interactions, as well as simulating quantum mechanical objects, such as a molecule with cancer-fighting properties.

Today's quantum computers are expensive, hard to build and run, and lack error correction. However, they're already valuable tools for quantum mechanics and quantum computing research and development, as well as helping to solve problems in specific fields and serving as a platform for the development of new algorithms for future quantum computers.

Where Is the Quantum Computing Field Going?

The potential of quantum computing is so great that it's seen as not only a research and development tool and a business opportunity but also a national security and military issue. (The first digital computers were developed for codebreaking in World War II, so this is not a new kind of concern.) The future of quantum computing is both important and, as quantum mechanics itself might suggest, impossible to predict with much accuracy.

It's early days for quantum computing; the most advanced quantum computers today have dozens or a few hundred qubits and are expensive to create and run. They lack error-correction capabilities, so results are likely to be inexact. These computers are very useful for research and development and also help solve practical problems, with varying degrees of assistance from classical (mainstream) computers.

Future quantum computers are likely to have thousands or even millions of qubits, be much cheaper to create and run, and include built-in error correction. As the technology progresses in this direction — faster, cheaper, better — quantum computers will solve more and more problems.

As quantum computers improve, they'll be used to help with designing future generations of quantum computers. AI and machine learning will probably be part of this evolution as well. As these technologies are brought to bear, progress is likely to speed up further. (And some kind of rapid breakout of new quantum computing or quantum communications capability, under the control of a rival government or an ill-intentioned organization or individual, is the focus of much of the current concern around quantum computing.)

When Will Quantum Computing Become Commercially Practical?

A substantial number of large corporations are investing in quantum computing now. Valuing the longer-term effect of these early investments in real time is challenging.

The answer to when quantum computing will become broadly important is often given as "10 years or so," with the joke being that the answer has been the same for the last 10 or 20 years.

But the current pace of change and independent estimates from relatively conservative analysts do point to the early- to mid-2030s as the period when truly game-changing progress will manifest. If true, forward-looking individuals and organizations will do well to start getting involved now.

REMEMBER

In the 1970s, few computers were connected to each other; information sharing was accomplished using low-capacity storage media called floppy disks. In the 1980s, networking computers together (using wires!) became practical, and trade publications started proclaiming "The Year of the Network." It took a while, and "The Year of the Network" was proclaimed for about five or six years in a row, which was funny. But by the end of the 1980s, the majority of office computers were networked. Similarly, it's hard to predict the exact date when specific increments of progress will be made in quantum computing, but they are likely to appear sooner rather than later.

What Is the Coolest Application of Quantum Computing?

There's no "best" answer to which quantum computing application is the coolest, so here are answers from each of the authors. And you may have your own!

Floyd: "My favorite application of quantum computing is using qubits to model or simulate or both the behavior of quantum particles in materials development. Being able to create fundamentally new "stuff" at scale is not only an awe-inspiring accomplishment; it will also help humanity fight climate change, the challenge that I am most concerned with. And better materials will help us create better quantum computers, which will then help to discover even better materials, which will help us create even better quantum computers . . ."

whurley: "I am a bit of a contrarian in that I am willing to let the hardware development proceed as it needs to, while I put my major focus on algorithms and software. Quantum advantage is dependent on new quantum algorithms, and these algorithms tend to build on each other. Already, we're learning fundamental things about computation and information theory from quantum software, and that is so cool. I believe that we need an increased focus on algorithm development now, which will pay off many times over in the years ahead."

Where Will Quantum Computing Be the Most Disruptive?

Technology is most disruptive when multiple developments layer on top of each other, creating a new reality that was almost unimaginable just a few years earlier. An electric car that can serve as a driverless taxi, called using apps like Uber or Lyft, is a good example of several advances combining to create a new experience.

We don't want to try to call out any one thing. A lot of people in quantum computing have to put their shoulders to a lot of grindstones to create breakthroughs over, say, the next 5 to 10 years that will lead to the hugely innovative accomplishments of tomorrow, perhaps 10 to 20 years out.

How Long Until Shor's Algorithm Breaks RSA?

Today's RSA typically uses 1,024, 2,048, or 4,096 bits. Cracking the low end of that could happen within 10 years; we would not bet anything important on it taking longer than that. But RSA is extensible; simply adding more bits to it will give it additional years of life. That will buy the industry time to develop and implement new, theoretically unbreakable, quantum-safe encryption.

How Can You Use Quantum Computing in Manufacturing?

We could almost turn this question around: What is there about a manufacturing setting that could *not* benefit from quantum computing? For instance, tremendous scope exists for the use of AI and machine learning in manufacturing; those disciplines, in turn, are highly likely to benefit from integration with quantum computing. (See the next question.)

But the fundamental change is that quantum computing will lead us to new materials, so the world can expect to be manufacturing new things, made of new materials, using new machinery and tools. Perhaps all of that will often be controlled by a combination of quantum computing, AI, and machine learning.

Where Is the Overlap Between Quantum Computing and AI/ML?

Machine learning proceeds by the use of tight algorithmic loops that chew through data at one level, and then pass it to different algorithmic loops at a higher level, and so on, training a neural-type net that can tackle really difficult problems. Quantum computing is likely to speed up some of these loops by orders of magnitude, with results that are difficult to predict.

The answer for AI, the broader field that includes machine learning, is a bit more hand-wavy for now. The pot of gold at the end of the rainbow for AI is artificial general intelligence (AGI): a machine that can pass for human across a wide range of tasks.

People have been pursuing this goal using classical computers, but this is in some ways a terrible approach. Classical computers are three things that the human mind is not:

- **Deterministic:** Classical computers always yield the same results from the same inputs. The human mind doesn't do that.
- **Unemotional:** Classical computers lack intuition, feel, an attraction to beauty, and other capabilities and heuristics that are actually vital to human reasoning.
- **Limited:** If you ask a classical computer to choose the right answer from a complex situation with many interacting actors, it will start trying to precisely calculate the answer for each interaction — and it will grind to a halt as the number of factors under consideration rises. Humans can rapidly integrate their way to conclusions across a *lot* of variables.

Quantum computers are neither deterministic nor limited in processing complex interactions. They are not yet emotional, but emotion-like heuristics may be easier to process on a quantum computer than a classical one.

» Finding out where to get educated in quantum computing

» Learning how to get funding and build a business

Chapter 19
Ten Business Questions Answered

One of the authors (whurley) gives a lot of quantum computing talks at conferences and other events around the world. At the end of each talk, he makes time for a Q&A session with the audience. The preceding chapter features questions from those Q&A sessions that concern quantum computing technology. Here, we share questions and answers from these sessions about the business side of quantum computing.

How Can I Evaluate the Market for a New Company, Product, or Service?

Quantum computing is about as innovative as it gets, so evaluating market spaces is hard. Here are a few questions for entrepreneurs to consider:

» Do you or your organization have experience and a reputation in quantum computing or in the broader field of high-performance computing (HPC)? If not, do you have adjacent experience and reputation that can you leverage?

» Can your business idea get traction in this early stage of the market?

- How many scarce technical people do you need to hire to develop your minimum viable product (MVP) and attract and support the first tranche of customers?
- What do you need to do up front — hiring staff, getting a product ready, months (or years) of effort, money spent — before you can start getting customers?
- How can you get involved in quantum computing now so you can hone your ideas and start connecting with people who can help you bring your idea to life?

How Do I Evaluate My Employer's Need to Be an Early Adopter (or Not)?

This question refers to intrapreneurship — taking entrepreneurial-type risks from within an existing company, even one that may usually be risk-averse. Here are a few questions to consider:

- What's your organization's usual risk appetite? That is, if you look at a half dozen competing companies or other companies similar to yours, does your organization rank high, medium, or low in pursuing emerging opportunities? Also, how favorably do people in your organization tend to react to new ideas?
- Are you one of the larger companies in your field? Larger companies usually have more resources for new opportunities, and both greater rewards if they pursue the new opportunities and greater risk if they don't.
- Does your company have quantum computing expertise already? Are any existing projects or efforts underway, even informal ones?
- What are the risks to your organization if a competitor becomes known as the leader in quantum computing adoption within your competitive set?
- What are the potential benefits to your organization if you become known as the leader in quantum computing adoption within your competitive set?

You can spot a few assumptions baked into these questions. First, people are usually more urgently motivated by fear (risk of loss or threats from competitors) than by hope (the opportunity for gain). No one wants to end up like Blockbuster Video after Netflix got traction: road kill on the information superhighway.

The opportunity for gain is inspiring but less motivating than fear. As Machiavelli put it, "It is better to be feared than loved," and this applies to business opportunities as well as statecraft.

If we were presenting to upper management, looking to sell them on a quantum computing initiative, we'd draw on both hope and fear, but lean on fear first. Paint the picture where your company gets run over by the competition, and then pivot to painting the picture where your company is the one that races ahead.

What Roles and Jobs Are Needed in the Current Stage of Development?

Today the need is great for technical people who can help move quantum computing forward: electrical engineers, physicists, and low-level software developers (especially developers who can work productively alongside electrical engineers and physicists). The more experience you have in fields related to these, the better. But if you are in one of these professions and have a solid track record in your current work, there's a good chance you can self-educate and network your way into quantum computing.

It's harder for people in non-technical fields — business and marketing, for example. How do you gain experience in quantum computing when you don't already have experience in quantum computing?

Luckily, that information superhighway thingy we mentioned in the preceding question is available now. You can use online resources like those described in Chapter 16 to connect, learn, and contribute. Helping others today means others may want you to help them tomorrow.

The partnership between this book's coauthors is an example. Years ago, one of us (Smith) was visiting London and nearly emptied Foyle's bookshop of books on quantum computing. (The Brits are leaders in quantum computing and tend to love eccentric scientist-type stuff. It was an Englishwoman who created Dr. Frankenstein and his monster, after all.) Smith then wrote to a senior editor at Wiley, the publisher of this book, asking if they needed an introductory book about quantum computing.

Wiley first reached out to the other of us (whurley), who was already well-known in the field as the founder and CEO of Strangeworks, about writing a book. whurley was too busy to do all of it himself; Smith already knew how to write *Dummies* books; and the team was off and running.

What Background Is Needed to Learn Quantum Computer Coding?

Software development for quantum computing requires skill in specific languages such as Python and C#, as well as the ability to pick up new languages and new concepts quickly. Having a math background is also important; much of quantum computing is about matrix manipulation, which is "only" algebra, not even calculus.

TECHNICAL STUFF

Because of the nature of quantum computing programming, we like the idea of reviving some older, math-oriented, "close to the bare metal" languages like APL (an acronym for "a programming language") and Fortran (a portmanteau of the words "formula translation") for quantum computing.

It's also good to have a familiarity with quantum mechanics principles. Some developers are allergic to the more alarming implications of quantum mechanics, such as the possible role of consciousness in, well, existence. "Shut up and code" is their mantra. But we find that the type of big-picture thinking needed to better understand quantum mechanics opens up people's minds, so that when they do "shut up and code," they are quick to discover and implement creative new approaches.

If you don't already have relevant programming experience, a math background, and quantum mechanics knowledge, learning quantum computing is harder. We're sure there will be people who succeed without having these elements in place, but doing so will require ascending a few steep learning curves.

What Advice Can You Give to First-Timers?

As a poet once wrote, you can get by with a little help from your friends. Reach out and get involved — online first, then in person, and through existing connections. Find ways to help others as they seek to make progress in this field and you're likely to become a valued resource, making it easier to join with others who have skills that you lack.

What University Programs Would You Recommend?

Award winners at shows like the Oscars often mention their fear of leaving someone out, and now we know what they mean. But what the heck — here are a half dozen programs from among the many leaders worldwide:

- The University of Waterloo in Canada has an institute for quantum computing that was founded in 2002 (!) and now has hundreds of leading researchers. They work hard on commercialization too.
- The University of Oxford is an early leader in quantum computing, with contributions going back to the 1980s. They're interested in seeing that quantum computing benefits society as a whole.
- In the US, Harvard, UC Berkeley, and MIT are among the pioneers. MIT's work includes quantum information theory right alongside quantum computing.
- The University of Sydney is a leader in quantum computing, emphasizing both the quantum mechanical roots of quantum computing and entrepreneurship.

We've left out many other important players, but these universities should certainly be near the top of anyone's list. See the next chapter for more.

Who Is the Current Leading Developer of Quantum Computing?

If there were a Coke versus Pepsi face-off on the business side of quantum computing, it would be between IBM, the century-old company that was once the undisputed leader in mainframe computing — almost the only kind, at that time — and Google, arguably the leader in computing today. (Apple, which is certainly a Google competitor, is regarded more as a consumer electronics company these days.)

Google is the company most closely identified with quantum computing. They got the biggest news spike in the field to date in 2019 when they announced the first widely credited instance of quantum supremacy: a computing operation that completed far faster on a quantum computer (built, of course, by Google) than it could run on a classical computer.

And IBM was the company that called them out as incorrect, claiming that the accomplishment was on the trivial side and not provably a great deal faster than classical computers could achieve. Controversy continues, but the back and forth reflects the competition between Google and IBM for leadership in quantum computing.

You might wonder why these two companies lead. Google has an obvious reason to want leadership in quantum computing: With most of their revenue coming from internet searches and search advertising, quantum computing is of existential importance to them. To search one million items for a single result, using classical computing, is going to need an average of 500,000 queries (1/2 *n*); the same search takes 1,000 searches, on average, with quantum computing (sqrt *n*). The advantage in this example is 50x. And Google does many searches where *n* is much larger than a million, making the quantum advantage even more pronounced.

Google also does a lot of searching and matching to place Adwords ads next to search results. And Google's business is currently being challenged by ChatGPT, Microsoft (an investor in ChatGPT), and Microsoft's Bing search engine. So if Google gets quantum-computing-powered search capability first, they will be able to provide a much better service while saving many billions of dollars. If a competitor gets the same capability first, on the other hand, they'll have a huge competitive leg up on Google.

Today's IBM, by contrast, is a service provider to many other businesses. Their business will become much stronger if they establish a worldwide leadership position in quantum computing, so the task is worth a lot of effort. Quantum computing is not an existential threat to IBM in the same way that it is to Google, and IBM is also a smaller company with one-tenth the market capitalization of Google and fewer resources to invest in something as cutting-edge as quantum computing. Yet IBM is making impressive strides toward leadership in quantum computing.

What Should I Do if I Have an Idea for a Startup?

Most of us who think of creating a company tend to place a high value on original ideas, but investors don't value ideas that highly. Instead, investors tend to place a high value on the proven ability to bring new ideas to fruition. For instance, Google got funding only after they had a working version of their search engine; Apple got funding only after they achieved initial success with the Apple I and were ready to manufacture the Apple II.

So a good idea is not usually enough. A working prototype and feedback from prospective — or, better yet, actual — customers, or even some early revenues, are needed. And it helps if you can show that you have the connections and experience to bring your idea to life. (The founders of Google were Stanford computer science students, surrounded by advisors and potential employees; the Apple founders were already well-known among electronics hobbyists.)

If you meet those criteria now, or if you want to roll up your sleeves and get to where you do meet them, start reaching out, find people who share your interests and concerns, and start recruiting allies. Only when you have something worth showing to others are you likely to get angel investment and help.

What Habits Have Helped You in Your Career?

Successful entrepreneurs tend to share certain traits, but there are a lot of successful people who don't have those traits. Also, entrepreneurship is a team sport; sometimes, for example, "people people" who lack advanced technical skills are able to recruit engineering rock stars to go seek investment with.

A few habits or approaches that we favor are to work on something you're passionate about; develop a reputation as an expert in your area (whether technical skills or running a business); get to know your proposed customer base really well; and become a capable self-promoter. If you achieve these things, you're likely to accomplish a lot, whether that's through entrepreneurship, employment, or some other means.

What Are Your Biggest Lessons Learned?

Both of us are past 50 and have been in tech for a long time, which gives us plenty of scope for successes, failures, and lessons learned. Here are a few highlights of the lessons we've learned (mostly derived from personal lowlights):

- **Focus first on your best skills.** Most of us have one or a few unique gifts to give to the world. Honing in on and sharing those core skills is usually going to lead to success.

- **Stretch your wings.** Jumping from one end of the checkerboard to the other rarely works. But hopping over a square or two often brings just the change of viewpoint that you need.
- **Stay connected.** Don't network mindlessly, but try to steadily get and keep connections to the people you respect the most and enjoy interacting with the most.
- **Listen to your gut.** Few people have gotten into quantum computing, for instance, because a solid cost-benefit analysis indicated high odds of success if they were to do so. Up until now, anyway, the field is still largely a passion project, but not many of us who dived in early would want to get out of the pool and towel off right about now.

» Searching for qubits in Harvard Yard

» Enjoying innovation in the land down under

Chapter 20
Ten University Research Programs

This chapter describes ten of the top university research programs for quantum computing. Reasons for inclusion in this list include past meritorious service, such as early contributions to the topic, and more recent contributions and renown. We've arranged this list with consideration both to how early an institution's contributions have been and some ordering by language and geography. So please consider this a sampler from among many strong contributors, rather than a definitive list.

The key undergraduate majors for those interested in quantum computing are computer science and physics, in both cases with an emphasis on the relevant mathematics. Computer engineering is also applicable, but the hardware for quantum computing may settle down into one or a few standardized architectures, lessening the need for additional workers, sooner than the algorithms and other computer software required.

At the graduate level, physics undergraduates can move into doctoral programs in computer science, and vice versa. And simply studying mathematics at any level is never, in our opinion, a bad idea. We studied information systems and statistics (Smith) and electronic music (whurley) in college; one of us (Smith) has two degrees, and one of us (whurley) has none. So when it comes to our respect for math, we're speaking more from aspiration than experience.

Also, quantum computing research and education are experiencing tremendous growth, with many institutions adding new programs or expanding existing ones. (Forty universities now offer graduate studies in the UK alone, and most of those programs are new.)

So, for the next decade or two, both the strength of the education one might receive and the opportunities, prestige, and connections that will accrue to completing a given program in quantum computing are likely to shift more rapidly than in most other fields. Caveat scholaris — let the aspiring scholar beware.

WARNING

This is not a complete list, nor is it a ranking! For instance, we stayed with schools in English-speaking areas to reduce complexity for readers wanting to focus on that language. And we went geographically, from Oxford/UK (in memoriam Alan Turing) to the US (West Coast to East Coast), Canada, and Australia, rather than ranking schools in some kind of order of merit. And finally, we confess to an anti-recency bias; we believe that it's too early to pick winners from among the many schools that have made big investments in quantum computing education since 2020 or so.

University of Oxford, UK

Oxford is an early leader in computing and quantum computing, with David Deutsch and Artur Ekert of Oxford both making key contributions, as mentioned in Chapter 4. The first working quantum computer using nuclear magnetic resonance was demonstrated at Oxford and the University of York. Oxford continues in a leading position today, developing superconducting qubits and contributing strongly to both theory and experimentation.

University of California, Berkeley

Cal Berkeley is another early leader in quantum computing, with Paul Benioff of Cal helping to stoke initial interest in the field in the early 1980s, as mentioned in Chapter 4. The university hosts the Berkeley Center for Quantum Information and

Computation and has strong ties to Lawrence Livermore Labs, which has added a quantum-computing-specific lab. John Clauser, who shared the 2022 Nobel Prize in Physics for work on entanglement, did most of the relevant work at Cal.

Stanford University

Stanford made early contributions to the first working quantum computers and to the implementation of Shor's algorithm for factoring large numbers. Stanford is also the initial home of Google, which shares business leadership in quantum computing with IBM. Recent work at Stanford has led to the creation of a time crystal, a new form of matter that pushes hard up against the second law of thermodynamics, using a Google quantum computer.

Cal Tech

The California Institute of Technology was home to the late Richard Feynman, the first person to conceive of quantum computing for use in simulating quantum mechanics, as mentioned in Chapter 4. CalTech is a leader in theoretical physics, including theoretical work in quantum computing, and their schools include the Institute for Quantum Information and Matter.

Massachusetts Institute of Technology

MIT hosted the first conference on quantum computing, as mentioned in Chapter 4, and has made ongoing contributions to quantum computing. As with CalTech, MIT's strength in theoretical physics has translated into work in quantum computing. MIT has a focus on superconducting qubits and combines quantum information (QI) theory and quantum computing in an area of study called QI/QC. It is also the academic home of Peter Shor of Shor's algorithm fame.

Harvard University

Harvard researchers have contributed to early advances in quantum information and the university hosts the Harvard Quantum Initiative. Their work includes quantum computers based on neutral atoms (as opposed to ionized atoms), quantum networking, and quantum sensing.

University of Chicago

Scientists at the University of Chicago are leaders in the study of superconducting-based quantum computers as well as neutral atom systems and recently contributed to extending the length of coherence for qubits. The university is home to the Chicago Quantum Exchange, which takes quantum information as its starting point.

WHO'S ON FIRST?

"Who's on first?" is a comedy routine by the team of Abbott & Costello, initially broadcast on radio in 1938. (There was no television, and certainly no internet, that far back in the day; the telegraph was still a big deal, though.) The routine described a confusion betwixt two broadcasters when a ballplayer named Who was standing on first base, another named What occupied second base, and so on. People often say "Who's on first?" to communicate general confusion.

In that spirit, as we went to press (in mid-2023), IBM launched a $100 million partnership for academic help in their five-year mission to explore strange new worlds . . . whoops, we lost focus there. We mean their ten-year mission to develop a future 100,000-qubit quantum computer. As partners, they chose the University of Tokyo and the University of Chicago — only one of which, the University of Chicago, is on our list.

We share this to confess our inability to create a definitive list of top quantum computing university programs and to urge you to use our list as a starting point, not an exclusive compendium.

University of Maryland

The University of Maryland is the leading quantum research university closest to Washington, DC, the base for the US National Institute of Standards (NIST), which plays a leading role in quantum computing, the Pentagon, and other US federal government agencies. The university is partnered with NIST in the Joint Quantum Institute.

University of Waterloo

The University of Waterloo in the Canadian province of Ontario bet big on quantum computing early by establishing the Institute for Quantum Computing in 2002, and now employs roughly 300 researchers. They have since contributed to key developments in the field, have a focus on superconducting qubits, and emphasize commercialization of the technology.

University of New South Wales, Sydney

Australia is a hotbed of quantum computing progress, and UNSW Sydney is central to this. The school recently led the creation of "hot" qubits that work at 1.5K (1.5 degrees Celsius above absolute zero). This extremely cold temperature is still much warmer than the space-cold temperatures in the millikelvins (thousandths of a degree Kelvin) used by most quantum computers.

Index

B

C

D

E

F

G

H

I

J

K

L

O

P

Q

R

S

T

U

V

W

X

Y

Z

About the Authors

William Hurley, known as "whurley," is the founder and CEO of Strangeworks, a quantum computing startup based in Austin, Texas. His career includes technical leadership roles for companies such as Apple, IBM (where he was a Master Inventor), BMC Software, and Goldman Sachs; he founded two previous startups, Chaotic Moon (acquired by Accenture) and Honest Dollar (acquired by Goldman Sachs). whurley partnered with StackOverflow to create the world's largest community of quantum computing developers and founded the Quantum Standards Working Group at the Institute of Electrical and Electronics Engineers (IEEE). whurley is a regular contributor to TechCrunch and speaks worldwide on the potential power and impact of quantum computing and artificial intelligence. He is also co-author of *Quantum Computing For Babies.* He lives in Austin, Texas, with his wife and children.

Floyd Smith is director of product marketing for Onehouse, a Silicon Valley startup in Sunnyvale, California. His career includes marketing and technical roles for companies such as Apple, Electronic Arts, HSBC, Visa, and startups based in London, New Zealand, and Silicon Valley. Floyd is a prolific author and blogger, with book sales exceeding one million copies. His best-selling book to date, *Creating Web Pages For Dummies,* was an early success in introducing web page creation to hundreds of thousands of people; it was translated into more than half a dozen languages and went through nine editions over a period of 15 years. Floyd holds a BA from the University of San Francisco in Information Systems Management and an MSc from the London School of Economics in Information Systems. He lives on the coast in the San Francisco Bay Area near his two grown children and three granddaughters.

Dedication

This book is dedicated to my wife and sons for their unending love and support. Also in memory of my late friend, mentor, and business partner, Mike W. Erwin; you are sorely missed.

— whurley

This book is dedicated to my amazing granddaughters, Honour, Eva, and Mira.

— Floyd Smith

Authors' Acknowledgments

Quantum computing is very new, deeply technical, and wonderfully strange; capturing the current state of this new art in a book required an unusually intense and sustained effort. We would like to jointly acknowledge all the many people who worked to help. Among them are Susan Pink, our incredibly patient and hardworking editor; Andrea Delgado, our astute and incisive technical editor; Steven Hayes, executive editor at Wiley, who put the writing team together; Carole Jelen, Floyd's agent, without whom this effort could not have happened; and several people at Strangeworks who went above and beyond the call of duty to contribute. Dr. Andrew Ochoa is blazing new trails in the advancement and use of quantum computers; he contributed crucial technical insights at the beginning and end of the project. Idalia Friedson is making dreams come to life for Strangeworks' customers and contributed a uniquely valuable point of view, embodied in countless edits. And art director Casey Barthels and his team Nichole Majeske and Ada Onyiuke worked, at times around the clock, to set a very high bar for the quality and quantity of images in this book.

Publisher's Acknowledgments

Executive Editor: Steve Hayes

Project Editor: Susan Pink

Copy Editor: Susan Pink

Technical Editor: Andrea Delgado

Proofreader: Debbye Butler

Production Editor: Tamilmani Varadharaj

Cover Image: Courtesy of Strangeworks, Inc., created by Ada Onyiuke

PERSONAL ENRICHMENT

9781119187790
USA $26.00
CAN $31.99
UK £19.99

9781119179030
USA $21.99
CAN $25.99
UK £16.99

9781119293354
USA $24.99
CAN $29.99
UK £17.99

9781119293347
USA $22.99
CAN $27.99
UK £16.99

9781119310068
USA $22.99
CAN $27.99
UK £16.99

9781119235606
USA $24.99
CAN $29.99
UK £17.99

9781119251163
USA $24.99
CAN $29.99
UK £17.99

9781119235491
USA $26.99
CAN $31.99
UK £19.99

9781119279952
USA $24.99
CAN $29.99
UK £17.99

9781119283133
USA $24.99
CAN $29.99
UK £17.99

9781119287117
USA $24.99
CAN $29.99
UK £16.99

9781119130246
USA $22.99
CAN $27.99
UK £16.99

PROFESSIONAL DEVELOPMENT

9781119311041
USA $24.99
CAN $29.99
UK £17.99

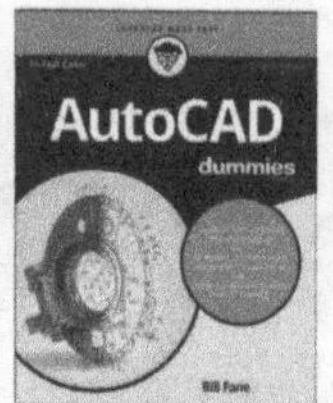

9781119255796
USA $39.99
CAN $47.99
UK £27.99

9781119293439
USA $26.99
CAN $31.99
UK £19.99

9781119281467
USA $26.99
CAN $31.99
UK £19.99

9781119280651
USA $29.99
CAN $35.99
UK £21.99

9781119251132
USA $24.99
CAN $29.99
UK £17.99

9781119310563
USA $34.00
CAN $41.99
UK £24.99

9781119181705
USA $29.99
CAN $35.99
UK £21.99

9781119263593
USA $26.99
CAN $31.99
UK £19.99

9781119257769
USA $29.99
CAN $35.99
UK £21.99

9781119293477
USA $26.99
CAN $31.99
UK £19.99

9781119265313
USA $24.99
CAN $29.99
UK £17.99

9781119239314
USA $29.99
CAN $35.99
UK £21.99

9781119293323
USA $29.99
CAN $35.99
UK £21.99